CREATIVE
SOUND
RECORDING
ON A
BUDGET

No. 2635
$19.95

CREATIVE SOUND RECORDING ON A BUDGET

DELTON T. HORN

TAB BOOKS Inc.
Blue Ridge Summit, PA 17214

FIRST EDITION

FIRST PRINTING

Copyright © 1987 by TAB BOOKS Inc.

Printed in the United States of America

Reproduction or publication of the content in any manner, without express permission of the publisher, is prohibited. No liability is assumed with respect to the use of the information herein.

Library of Congress Cataloging in Publication Data

Horn, Delton T.
    Creative sound recording on a budget.

    Includes index.
    1. Magnetic recorders and recording—Amateur's manuals. I. Title.
TK9967.H67   1986   621.389′3   86-5936
ISBN 0-8306-0635-1
ISBN 0-8306-2635-2 (pbk.)

Contents

	Introduction	ix
1	**The Nature of Sound**	1
	How We Hear—Amplitude—Frequency—Harmonics—Enharmonics—White Noise and Pink Noise	
2	**Tape Recorders**	23
	How Signals Are Recorded—Tape Tracks—Factors Determining Frequency Response—Record Bias—Tape Formats	
3	**DB and VU**	51
	Decibels—VU Meters—S/N Specifications	
4	**Microphones**	55
	Carbon Microphones—Crystal Microphones—Dynamic Microphones—Ribbon Microphones—Condenser Microphones—Electret Microphone—Microphone Pickup Patterns	
5	**Setting Up a Home Recording Studio**	67
	Selecting a Tape Machine—Monitor System—Impedance Matching—Studio Acoustics	
6	**Live Recording**	103
	Equipment for Live Recording—Setting Up	
7	**Care of Tape and Recorders**	109
	Tape Storage—Tape Handling—Cleaning Equipment and Supplies—Demagnetization—Lubrication—General Equipment Maintenance and Repairs	

8 Splicing 119
Reasons for Splicing—Making the Basic Splice—Splicing Blocks

9 Mixing 139
Sound-on-Sound—Sound-with-Sound—Four-Channel Decks—Multideck Approach—Mixers—Overdubbing—Editing in the Mix

10 Special Effects and Accessories 153
Equalization—Noise Reduction—Speed Effects—Reversing Tape Direction—Echo and Reverberation—Simple Electronic Effects

11 Tape Types 187
Reference Standards—Tape Sensitivity—Tape Noise—Maximum Modulation Level—Distortion Level—Mechanical Properties—Magnetic Properties—The Right Tape for Your Recorder—Watch Out for Bargains—Tape Formulations—Summary

Index 207

CREATIVE SOUND RECORDING ON A BUDGET

Introduction

This book is intended to help you use your tape recorder for more than simply dubbing your record albums. You don't need a full-fledged, professional-level recording studio to try your hand at creative recording. A lot of interesting and exciting things can be done at almost any budget level.

No previous experience on the part of the reader is assumed in this book, but this work is not just for beginners. The long-time hobbyist or semipro can pick up plenty of useful tips that might not be available from other sources.

Whatever your interest in recording—recording musical groups, playing multiple parts yourself, radio/tape drama, or almost anything else—this book should get you well on your way to realizing your creativity with tape equipment. Straightforward techniques and many special effects are covered, with suggestions on how they can be adapted to suit your individual requirements.

For the hobbyist or the semiprofessional, creative recording techniques can be valuable. Why not make your "audio dreams" come true?

Chapter 1

The Nature of Sound

This book is about the creative use of tape recording equipment to record and play back various types of sound. For any but the most straightforward type of recording, it is essential to have at least some understanding of the nature of sound. This chapter gives a brief introduction to *acoustics*, the science of sound. Many readers may tend to skip this theoretical material and jump right ahead to the discussions of actual recording equipment and its practical uses. It is strongly advised that you read this chapter anyway. A good background in acoustics is important for any serious recordist. Many of the discussions in later chapters build upon the basics explained here. In addition, a firm grasp of the theoretical aspects will allow you to make better, more creative, and practical use of your tape recording equipment.

The subject of acoustics cannot be covered thoroughly in just one chapter. It is a complex and multidimensioned subject. Several thick volumes could be, and have been, written on this topic. This chapter will just look at the most basic fundamentals of acoustics.

HOW WE HEAR

Sound is a series of fluctuations of air pressure. To understand what this means, consider a taut string with both ends held stationary, but free to vibrate in the middle, as shown in Fig. 1-1. This is the situation with most stringed instruments, such as guitars or violins.

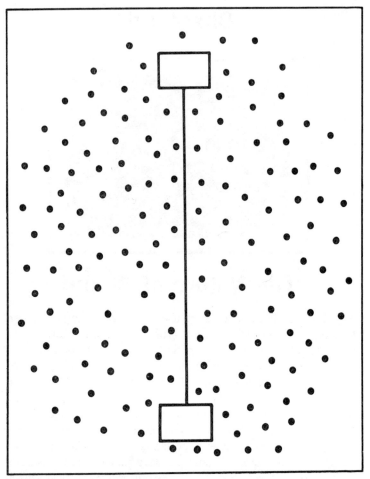

Fig. 1-1. A taut string with both ends held stationary can be used to illustrate how sound waves are generated.

Normally, the string is motionless, so the surrounding air molecules (represented by dots in the diagrams) are not disturbed in any way. They are more or less equally distributed on all sides of the string. This is called the *rest* or *null* position. (Note that only a few air molecules are shown in these illustrations for clarity.)

Now, suppose the string is stretched to the position shown in Fig. 1-2. At the instant of movement, the string takes up some of the space formerly occupied by the air molecules in front of it. The molecules on this side of the string are therefore compressed or forced closer together. Meanwhile, a small partial vacuum appears in the space the string has vacated. The air molecules on this side spread out to try to fill this "gap." When the string is moved in the other direction (Fig. 1-3), the same situation is created, but with the compression and expansion sides reversed.

If you pluck a taut string of this type, it will vibrate very rapidly, moving back and forth between the extreme positions shown in Figs. 1-2 and 1-3. Consequently, the air molecules on either side of the string are alternately compressed and expanded.

The movement of the vibrating string temporarily disturbs the

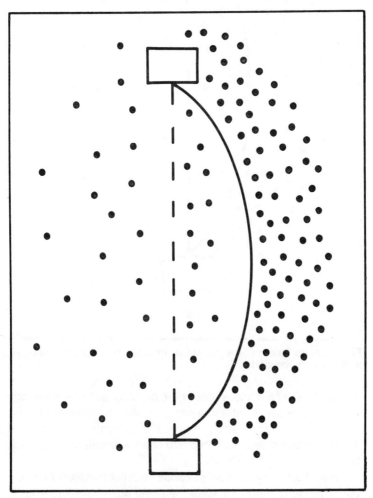

Fig. 1-2. At one extreme end of the string's motion, the air molecules on one side are compressed, while those on the opposite side are spread out.

3

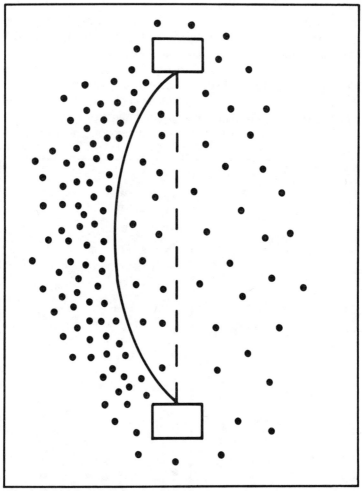

Fig. 1-3. At the other extreme end of the string's motion, the surrounding air molecules are reversed.

arrangement of the molecules in the air. Any disturbance of this type tends to move out and away from the source of the disturbance. The vibrating string moves the molecules nearest to it. These moving molecules bump into other molecules, further out, transmitting the disturbance.

Note that any individual molecule does not move very far, but the disturbance pattern is spread out over an ever-widening area, until the motivating force is used up. You can see the same sort of effect by tossing a pebble into a pool of water. A pattern of concentric waves spreads out from the point where the pebble struck the surface of the water (Fig. 1-4).

The pattern does not continue indefinitely. Energy is used up transmitting the pattern from molecule to molecule. As the original motivating energy that set the pattern into motion is depleted, the outermost waves grow weaker and weaker. The waves furthest from the center of disturbance are obviously the weakest. They have received less of the original motivating energy, and they are spread out over a larger area.

The air around a vibrating string behaves in essentially the same way as the water around the point where a pebble is tossed into it. Alternating waves of contraction and expansion spread outward from the vibrating surface (center of disturbance). These alternating changes of air pressure striking the eardrum are translated by the brain into the sensation of sound. The whole process is illustrated in Fig. 1-5.

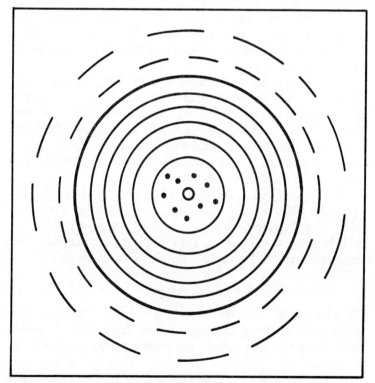

Fig. 1-4. Sound waves move out in growing concentric circles like the waves created when a pebble is thrown into water.

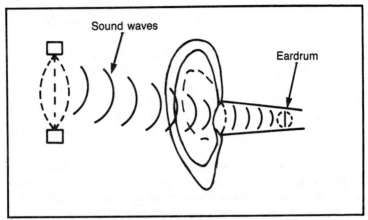

Fig. 1-5. Alternating changes in air pressure strike the eardrum and are translated into sensations of sound by the brain.

All sounds are produced in precisely this manner, regardless of their source. Some physical surface must be set into vibration to cause the waves of pressure disturbance. The vibrating string is the easiest to visualize, but all sound sources function via the same principles.

Air pressure may be disturbed by a vibrating string, a vibrating sheet or membrane (a drum head), a column of air forced through a tube (wind instruments), or any similar type of device. Even pure electronically generated sound signals must be converted into physical vibrations at some point in order to be perceived as sound. Usually this is done by causing the cone of a speaker to vibrate in step with the signal voltage.

To be perceived as sound the vibrations must be continuous over some finite period of time because one vibration is too brief to be heard. The vibration must also follow a more or less regular, periodic pattern of some kind. The rate of the vibrations is another important factor in the audibility of the disturbance pattern.

AMPLITUDE

The more severe the force causing the vibrations, the farther the string (or other vibrating surface) moves. The greater the movement of the vibrating surface, the greater the disturbance of air pressure.

The amount of air pressure disturbance is referred to as the *amplitude* of the signal. A large disturbance pattern is a *high-*

amplitude signal, and a small disturbance is a *low-amplitude signal.*

Amplitude is discussed in two basic ways. *Instantaneous amplitude* refers to a given point at a specific moment in time. Instantaneous amplitude fluctuates throughout the vibration/disturbance cycle. If you monitor one group of air molecules in Figs. 1-1 through 1-3, you will see that as the string vibrates, the molecules increasingly compress until a peak compression point is reached. Then they begin to spread out again, passing through their rest state, until an expansion peak is achieved.

Alternatively, the overall or *average volume* of the disturbance pattern as a whole can be studied. There are a number of ways such an average might be taken, but do not worry about the specific methods at this point.

Figure 1-6 shows a graph of nine instantaneous amplitude points within a single vibration cycle. The instantaneous amplitude is graphed against time. Look at the instantaneous pressure levels at one fixed location with respect to the string. Assume that you are in front of the string. The air pressure levels behind the string are an exact mirror image of this graph. (When one goes up, the other goes down, and vice versa.)

At the beginning of the cycle, the string starts out momentarily in its normal, rest position. The air pressure at this instant is neither increased nor decreased from the null condition, so the instantaneous amplitude at this point is zero. This point is labelled "A" on the graph.

Now the string starts to move forward, increasing the air pressure in front of it. (The pressure behind the string is reduced by

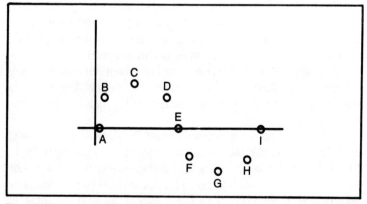

Fig. 1-6. The instantaneous amplitude is different at various points in the vibration cycle.

a like amount.) The increase in air pressure is due to the air molecules being forced closer together (compression). This is marked point "B" in Fig. 1-6.

The instantaneous amplitude continues to increase until it reaches its maximum level (point "C"). The string can't move any farther forward, but it still has energy to dissipate. It has to move back in the opposite direction, allowing the pressure in front to drop as the air molecules are allowed to spread out (point "D"). At some point, the string passes again through its rest position (point "E"). Here the instantaneous amplitude is again zero.

The string continues to move backwards, reducing the pressure in front of itself below the null state ("F"). We now have a negative instantaneous amplitude. The backwards motion continues until the maximum negative level is reached ("G"). Usually the maximum negative level is equal to the maximum positive level, making the pattern symmetrical. There are some exceptions, however.

Once the negative maximum level is reached, the string must again reverse direction and start to move forward again, increasing the pressure in front ("H"). When the string passes through the rest position again, a new cycle is started ("I").

The instantaneous amplitude is constantly changing as the string passes through these various points, never stopping at any one. By extending the graph to include an infinite number of instantaneous points, it ends up looking like Fig. 1-7. A graph of this type represents the waveshape of the signal. A complete vibration pattern, like the one shown in Fig. 1-7, is called a *cycle*.

There are many different waveshapes, which are produced under various circumstances. The simplest pattern is the one shown in Fig. 1-7 that is known as a *sine wave*. Other waveshapes and their characteristics are discussed later in this chapter.

A sound can easily be represented by a continuously changing (ac) electrical voltage. The same graphs would apply, but the points would represent instantaneous voltage rather than instantaneous amplitude.

The sine wave is symmetrical around the horizontal (amplitude) axis. In other words, the maximum positive instantaneous amplitude equals the maximum negative instantaneous amplitude. The sine wave is also symmetrical along the vertical (time) axis. The second half of the cycle is an exact mirror image of the first.

To discuss the overall amplitude of the waveform, rather than a series of instantaneous amplitudes, some kind of average of in-

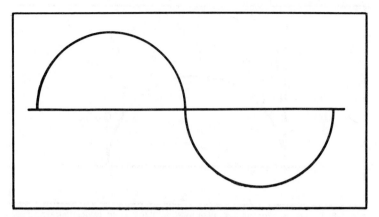

Fig. 1-7. The simplest waveshape is the sine wave.

stantaneous amplitudes must be taken throughout the cycle. Unfortunately, the vertical symmetry means that the average of the complete cycle always works out to zero. The negative half-cycle always exactly cancels out the positive half-cycle. Clearly this is not a very useful result.

To get a meaningful average, only one half-cycle is considered rather than a complete cycle. There are different ways in which the average may be taken. For audio waveshapes, the *RMS (root mean square)* method is generally used. Because the mathematics involved in calculating RMS values can be fairly complex, a straightforward mathematical average is used here.

To calculate the average, take several equally spaced instantaneous amplitudes in a single half-cycle. In Fig. 1-8, the instantaneous amplitude is sampled seven times in the course of the half-cycle. Assume the marked instantaneous amplitudes have the following values:

$$A = 0.0$$
$$B = 1.2$$
$$C = 2.7$$
$$D = 4.1$$
$$E = 2.7$$
$$F = 1.2$$
$$G = 0.0$$

The mathematical average for these values works out to 1.7 units. The units in this example are arbitrarily defined.

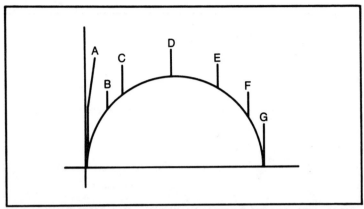

Fig. 1-8. To get the average overall amplitude of the wave, take several equally spaced instantaneous amplitudes in a single half-cycle.

Acoustic amplitude, or sound level, is often measured with a unit called a *bel* (abbreviated B). This unit is really too large for most practical applications, so the *decibel* (dB) is more commonly used. Ten decibels equal one bel.

The bel and the decibel are relative terms. They are not absolute units like the volt. To say that sound A has a level of 25 dB is completely meaningless. The bel and the decibel are comparative units, therefore, a reference standard must be specified. The correct form would be to say that sound A is 25 dB greater than sound B.

Bels and decibels are also nonlinear units. A doubling of amplitude is a change of 6 dB; that is, if sound A is 6 dB, referenced to sound B, then sound A will have twice the amplitude of sound B.

The decibel system might seem to be rather awkward to use, and to some extent, it is. There are a number of significant reasons why it is preferable to using an absolute unit of measurement such as a volt. The primary reason for using the decibel system is that the ear perceives amplitude in an exponential, rather than a linear manner.

An absolute unit of measurement is, by definition, linear; that is, two units are twice as large as one unit, but only half as large as four units. A literal doubling of the air pressure vibrations in a sound does not result in twice the perceived volume, however. In fact, such a doubling would barely be an audible change at all.

For one sound to be heard as twice as loud as another, it must move about ten times as much air. (This is equivalent to about a

6 dB difference.) Because of the way the human ear responds to changes in amplitude, an absolute measurement system would actually be more awkward to use. (There are some absolute measurement units for sound waves, but their use is generally limited to extremely technical situations.)

Another reason for using the decibel system involves the way the human ear and brain block background noise. A sharp hand clap is perceived considerably louder in a library than in a busy factory even though the absolute amplitude (air movement) may be identical in both cases.

True absolute silence does not occur naturally on Earth, so there is always some background noise in any normal environment. By using the environmental background noise level as a reference point, the amplitude of a sound can be described in a way that closely reflects the way it is perceived. The relative nature of the decibel system permits this type of referencing.

The decibel system is of great importance in sound recording work. It will be discussed again later in this book.

FREQUENCY

A single vibration cycle does not produce an audible sound. It is too short for the ear to perceive. Sound consists of a series of periodic vibrations. Generally, adjacent vibrations occur at a regular rate (frequency) and at similar average amplitude levels. This produces a recognizable sense of pitch.

The vibration rate is measured in the number of *cycles per second*. This is often abbreviated as *cps*. Another word for the same thing is *hertz*, or *Hz*. Hertz and cps mean exactly the same thing; they are interchangeable. In recent times, hertz (Hz) has become the preferred term, but cps is still occasionally used.

For higher frequencies, the unit of measurement is the *kilohertz* (*kHz*). One kilohertz is equal to one thousand hertz. For example, 1700 Hz = 1.7 kHz. Similarly, 22.4 kHz = 22,400 Hz. You might also occasionally come across the term *kilocycle* (*Kcs*). This is exactly the same as the kilohertz.

One million hertz (one thousand kilohertz) may be expressed as one *megahertz* (*MHz*). This book doesn't deal with megahertz values because such high frequencies are far beyond the audible range.

The human ear is sensitive only to a relatively narrow band of frequencies. It also tends to be more sensitive to changes in frequency than to changes in amplitude. The exact range varies with the individual and, to some extent, the listening conditions. The lower limit of audibility is typically between 20 and 60 Hz. The upper limit of audibility is generally considered to be between 15,000 and 20,000 Hz (15 to 20 kHz). The audible range, especially the upper end, decreases with the listener's age. This deterioration of the audible frequency range starts in the early twenties for most people.

Incidentally, this limit to the audible frequency range is the secret to those "silent" dog whistles that dogs can hear, but humans can't. Dogs can hear higher frequencies than people. These "silent" whistles simply emit vibrations at about 25,000 Hz (25 kHz).

Frequencies above the human audible range are called *ultrasonic*. By the same token, frequencies too low to be audible are called *subsonic*.

For musical purposes, the audible range of frequencies is divided into a number of *octaves*. Doubling a frequency raises it one octave. For example, 880 Hz is exactly one octave higher than 440 Hz. 1760 Hz is one octave higher than 880 Hz and two octaves higher than 440 Hz. The full nominal audible range for humans

Table 1-1. Standard Frequencies for the Traditional Western Music Scale.

C	16.4	32.7	65.4	130.8	
C# (Db)	17.3	34.6	69.3	138.6	
D	18.4	36.7	73.4	146.8	
D# (Eb)	19.4	38.9	77.8	155.6	
E	20.6	41.2	82.4	164.8	
F	21.8	43.7	87.3	174.6	
F# (Gb)	23.1	46.2	92.5	185.0	
G	24.5	49.0	98.0	196.0	
G# (Ab)	26.0	51.9	103.8	207.7	
A	27.5	55.0	110.0	220.0	
A# (Bb)	29.1	58.3	116.5	233.1	
B	30.9	61.7	123.5	246.9	
C	261.6	523.3	1046.5	2093.0	4186.0
C# (Db)	277.2	554.4	1108.7	2217.5	4434.9
D	293.7	587.3	1174.7	2349.3	4698.6
D# (Eb)	311.1	622.3	1244.5	2489.0	4978.0
E	329.6	659.3	1318.5	2637.0	5274.0
F	349.2	698.5	1396.9	2793.8	5587.7
F# (Gb)	370.0	740.0	1480.0	2960.0	5919.9
G	392.0	784.0	1568.0	3136.0	6271.9
G# (Ab)	415.3	830.6	1661.2	3322.4	6644.9
A	440.0	880.0	1760.0	3520.0	7040.0
A# (Bb)	466.2	932.3	1864.7	3729.3	7458.6
B	493.9	987.8	1975.5	3951.1	7902.1

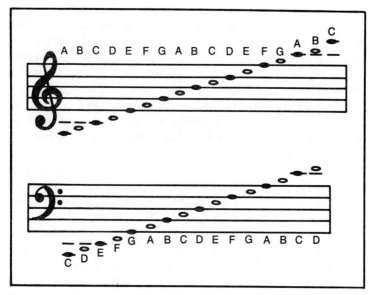

Fig. 1-9. The frequencies of the traditional Western scale are indicated by standardized symbols in a musical score.

covers about ten octaves. In traditional Western music, each octave is further subdivided into 12 specific *tones*. Each tone or note is higher than the previous one by a factor of the twelfth root of 2 ($^{12}\sqrt{2}$).

Table 1-1 lists the standard frequencies used for the traditional Western musical scale. Figure 1-9 illustrates how these frequencies are indicated in a musical score.

Other scales are also possible. Much oriental music is composed using a *pentatonic* (five notes per octave) scale. A twentieth century composer named Harry Partch devised a system with 43 discrete notes within each octave. This scale approaches the limit of human perception in distinguishing between adjacent frequencies. Naturally, Partch had to design and build all of his own instruments to utilize this unique extended musical scale.

By varying pitches, melodies are formed. Without some way to control the frequency of sounds, music would not be possible.

HARMONICS

The two most important aspects of a sound are the frequency and the amplitude, but there are certainly other significant factors. Different sound sources might produce the same frequency at the same amplitude, and yet sound completely different.

One of the most important factors in determining the nature of the sound is the waveshape. The sine wave (Fig. 1-7) is the simplest possible waveshape. It is the only truly pure signal, consisting of only a single frequency component. The only frequency involved is the basic repetition rate of the complete cycle.

Such a pure signal does not occur naturally. In fact, it can be quite difficult to accomplish artificially. This is just as well because a pure sine wave is a very piercing sound that can become extremely irritating very quickly.

Most sounds in the real world have more complex waveshapes comprised of multiple frequency components. The basic cycle repetition rate of the waveshape is the *fundamental frequency*. If the sound has a strong sense of pitch, the perceived pitch (tone) usually (but not always) is equal to the fundamental frequency. Many sounds do not have a strong sense of pitch. The fundamental frequency of such sounds is hard to recognize audibly.

Virtually all sounds also include frequency components higher than the fundamental frequency. These higher frequency components are called *overtones*. If there are any frequency components lower than the fundamental, they are called *undertones*. Undertones are rather rare but do occur occasionally.

There are two types of overtones. If the overtone frequency is an exact integer multiple (2X, 3X, 4X, etc.) of the fundamental frequency, the overtone is called a *harmonic*. If the overtone frequency is not an exact integer multiple of the fundamental, it is called an *enharmonic*.

The second harmonic is equal to two times the fundamental, the third harmonic is equal to three times the fundamental, and so forth. Notice that there is no first harmonic, since one times the fundamental would be identical to the fundamental.

Theoretically, any sound can be created by combining sine waves at the appropriate frequencies and relative amplitudes. Each sine wave is a single frequency component. Instead of being heard as a collection of simultaneous sine waves, they blend into a single more complex waveform with its own unique tonal quality.

The sine wave frequency components affect the waveshape. Figure 1-10 illustrates what happens when a fundamental is combined with its second harmonic. This process can be extended to include any desired number of frequency components.

To give you a clearer idea of how harmonics work, consider a complex sound that contains all of the harmonics up to the tenth.

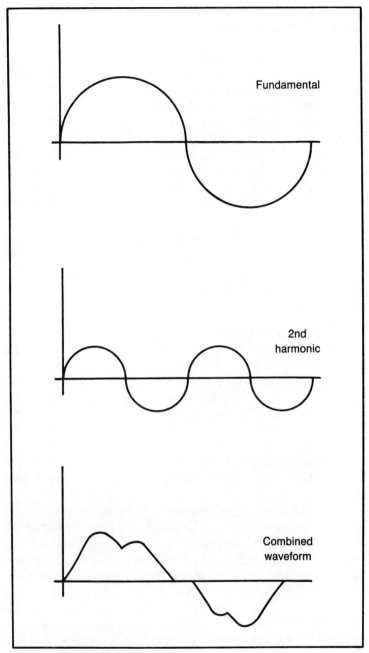

Fig. 1-10. The waveshape is altered when the second harmonic is added to the fundamental.

If the fundamental frequency is 100 Hz, the signal contains the following frequency components:

100 Hz	FUNDAMENTAL	
200 Hz	SECOND HARMONIC	(2 × 100)
300 Hz	THIRD HARMONIC	(3 × 100)
400 Hz	FOURTH HARMONIC	(4 × 100)
500 Hz	FIFTH HARMONIC	(5 × 100)
600 Hz	SIXTH HARMONIC	(6 × 100)
700 Hz	SEVENTH HARMONIC	(7 × 100)
800 Hz	EIGHTH HARMONIC	(8 × 100)
900 Hz	NINTH HARMONIC	(9 × 100)
1000 Hz	TENTH HARMONIC	(10 × 100)

If we change the fundamental frequency, the frequency of all the harmonics will also be changed by a like amount to maintain the same mathematical relationships. For example, here's what happens when we change the fundamental frequency in the signal just described to 225 Hz:

225 Hz	FUNDAMENTAL	
450 Hz	SECOND HARMONIC	(2 × 225)
675 Hz	THIRD HARMONIC	(3 × 225)
900 Hz	FOURTH HARMONIC	(4 × 225)
1125 Hz	FIFTH HARMONIC	(5 × 225)
1350 Hz	SIXTH HARMONIC	(6 × 225)
1575 Hz	SEVENTH HARMONIC	(7 × 225)
1800 Hz	EIGHTH HARMONIC	(8 × 225)
2025 Hz	NINTH HARMONIC	(9 × 225)
2250 Hz	TENTH HARMONIC	(10 × 225)

Not all harmonics are included in every waveshape. For instance, a square wave (Fig. 1-11) includes only the odd numbered harmonics (third, fifth, seventh, etc.). All of the even numbered harmonics (second, fourth, sixth, etc.) are omitted. The relative amplitude of each harmonic can be calculated with this simple formula:

$$A_h = (1/H) \times A_f$$

where A_h is the amplitude of the harmonic in question, H is the harmonic number, and A_f is the amplitude of the fundamental. Consequently, we get this pattern:

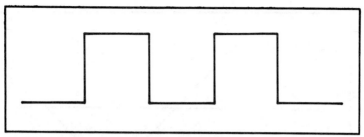

Fig. 1-11. A square wave includes a fundamental and all of the odd harmonics.

FUNDAMENTAL	100%	
SECOND HARMONIC	0%	
THIRD HARMONIC	33.33%	1/3
FOURTH HARMONIC	0%	
FIFTH HARMONIC	20%	1/5
SIXTH HARMONIC	0%	
SEVENTH HARMONIC	14.28%	1/7
EIGHTH HARMONIC	0%	
NINTH HARMONIC	11.11%	1/9
TENTH HARMONIC	0%	
ELEVENTH HARMONIC	9.09%	1/11
TWELFTH HARMONIC	0%	
THIRTEENTH HARMONIC	7.69%	1/13
FOURTEENTH HARMONIC	0%	
FIFTEENTH HARMONIC	6.06%	1/15

Notice that as the harmonics get higher, they get weaker.

The specific harmonics included in the signal affect the waveshape and, therefore, the tonal quality. The relative proportions of the frequency components also play a significant role. For example, consider the triangle wave that is shown in Fig. 1-12. This waveform includes the same frequency components found in the square wave just discussed. It looks and sounds quite different, however. This is because the harmonics in a triangle wave are much weaker than those in a square wave. Instead of multiplying the fundamental amplitude by the reciprocal of the harmonic number (1/H), for a triangle wave the harmonic amplitudes are found by multiplying the fundamental amplitude by the reciprocal of the square of the harmonic number (1/(H × H)):

FUNDAMENTAL	100%
SECOND HARMONIC	0%

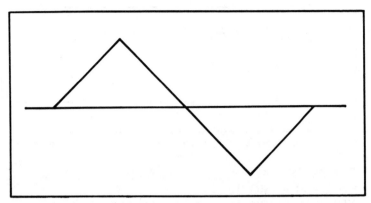

Fig. 1-12. A triangle wave has the same harmonic content as a square wave, but the harmonics are weaker.

THIRD HARMONIC	11.11%	1/9
FOURTH HARMONIC	0%	
FIFTH HARMONIC	4%	1/25
SIXTH HARMONIC	0%	
SEVENTH HARMONIC	2.04%	1/49
EIGHTH HARMONIC	0%	
NINTH HARMONIC	1.23%	1/81
TENTH HARMONIC	0%	
ELEVENTH HARMONIC	0.83%	1/121
TWELFTH HARMONIC	0%	
THIRTEENTH HARMONIC	0.5%	1/199
FOURTEENTH HARMONIC	0%	
FIFTEENTH HARMONIC	0.44%	1/225

You can see that the harmonics in a triangle wave are considerably weaker than in a square wave even though the frequency relationships are the same in both signals.

ENHARMONICS

Most natural sound sources produce signals with some enharmonic content. An enharmonic, as stated earlier, is an overtone (or occasionally an undertone) that has a frequency that is *not* an exact integer multiple of the fundamental frequency. For example, if the fundamental frequency is 200 Hz, and the signal includes a frequency component at 450 Hz, that frequency component is an enharmonic.

A small amount of enharmonic content adds richness and life

to a sound. This is why the much purer waveforms generated by electronic instruments often sound rather unnatural. (There are ways to get around this problem in producing electronic music, but they are beyond the scope of this book. If you are interested in electronic music, I would recommend that you read *The Beginner's Book Of Electronic Music* (TAB #1438).)

Because an enharmonic is not an exact integer multiple of the fundamental, the cycles do not begin and end at the same time (Fig. 1-13). This reduces the periodic repetition of the waveform. If the enharmonic content is strong, the sense of definite pitch is weakened and eventually lost altogether. If a metal garbage can is knocked over, the sound produced has a high enharmonic content.

Musical tones generally have strong harmonics and weak enharmonics. For unmusical sounds, the reverse is usually true.

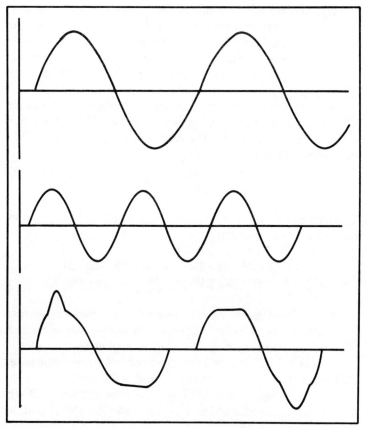

Fig. 1-13. Enharmonic overtones upset the repeating pattern of cycles.

WHITE NOISE AND PINK NOISE

There is one more special class of sounds to consider. This is noise. In acoustical and electronics work, noise has a specific meaning. Aunt Martha may call rock and roll "noise," but that's just her opinion. For our purposes, noise is a technical term with a precise meaning.

Noise is a nonperiodic (nonrepeating) signal with a large number of random (or pseudorandom) enharmonic frequency components. Noise does not have a fundamental frequency. The various frequency components appear and disappear irregularly.

There are two basic types of noise—white noise and pink noise. *White noise* is analogous to white light. White light is a combination of all visual colors. Similarly, white noise is a combination of all audible frequencies. At any given instant, any specific frequency has an equal chance of being present as any other specific frequency. Another way of saying this is that white noise contains an equal amount of energy per frequency. If you tune an FM receiver between stations, the hiss you hear is more or less white noise.

Pink noise is similar except that instead of equal energy per frequency, it features equal energy per octave. What's the difference?

An octave is a doubling of frequency. In the following example, while ignoring fractional frequencies such as 3.3 Hz, if you start with a frequency of 10 Hz, the next octave starts at 20 Hz. The first octave includes ten frequencies:

10 - 11 - 12 - 13 - 14 - 15 - 16 - 17 - 18 - 19

The next octave, however, extends from 20 Hz to 40 Hz, so it includes twenty frequencies:

21 - 22 - 23 - 24 - 25 - 26 - 27 - 28 - 29 - 30 -
31 - 32 - 33 - 34 - 35 - 36 - 37 - 38 - 39

Higher frequencies contain more discrete frequencies than lower frequencies; therefore, the upper frequencies are emphasized. As a result, white noise sounds fairly high-pitched.

In pink noise, frequencies in upper octaves are de-emphasized Pink noise sounds more "random" than white noise.

Pink noise can be derived from white noise by passing the signal through a low-pass filter. A *low-pass filter* is a circuit that, as its name suggests, passes low frequency components, but increasingly attenuates (reduces the amplitude of) higher frequency components.

Successfully reproducing white or pink noise is a good test of the frequency response of any sound system (amplifier, recorder, or whatever). A noise generator can also come in handy for the creative recordist in other ways. Many special effects can be produced by manipulating the output of a noise generator. Both white and pink noise can be useful for simulating percussion instruments (drums) and a number of nonmusical sound effects (rain falling, motors, wind, etc.)

Chapter 2

Tape Recorders

As indicated in Chapter 1, a sound waveform can be represented by a similarly shaped electrical waveform. Sound waves can be converted into varying voltages that can be stored as magnetic patterns on a strip of magnetic tape. Later, the magnetic patterns can be converted back into varying voltages and then back into sound waves again. The device that performs this near miracle is called a tape recorder. There are many different types of tape recorders, but they all work according to the same basic principles.

HOW SIGNALS ARE RECORDED

Recording tape is simply a long strip of plastic that has been coated with magnetically sensitive particles. Usually this is some type of ferric oxide, but newer technology uses other substances, such as chromium dioxide, to make tapes. This is especially true of tapes in the cassette format.

Record Head

On a blank (unrecorded) tape, the magnetic particles are arranged haphazardly (Fig. 2-1). If a strong magnetic field is placed across the tape, the magnetic particles line up in the magnetic field (Fig. 2-2). Lining up all of the particles in one direction doesn't really accomplish very much.

The magnetic field across the tape is driven by an ac (fluctuat-

Fig. 2-1. On a blank tape, the magnetic particles are arranged haphazardly.

Fig. 2-2. If a strong magnetic field is placed across the tape, the magnetic particles will line up in the magnetic field.

ing) signal—specifically, a varying voltage that represents the sound waves to be recorded. Consequently, the alignment of the magnetic particles on the tape corresponds to the varying voltage that corresponds to the instantaneous amplitude of the sound being recorded. This is illustrated in Fig. 2-3.

The fluctuating magnetic field is created by feeding the varying voltage signal to a device called a *record head*. The signal voltage is fed through a coil that causes a varying magnetic field. A simplified diagram of a record head's construction is shown in Fig. 2-4.

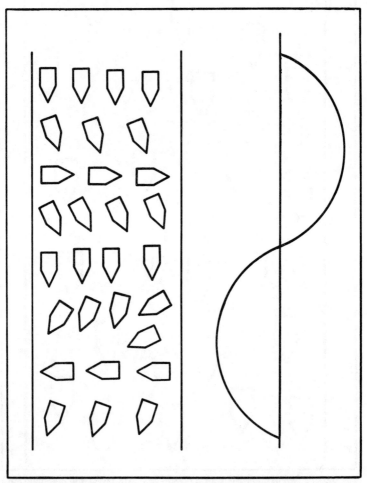

Fig. 2-3. The alignment of the magnetic particles on the tape corresponds to the instantaneous amplitude of the sound being recorded.

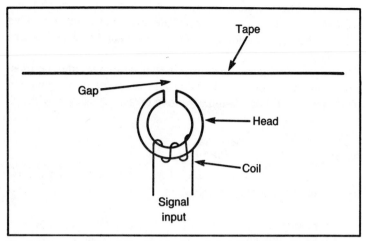

Fig. 2-4. In a record head, the signal voltage is fed through a coil, which generates a varying magnetic field.

Playback Head

Now that information representing the sound has been magnetically stored on the tape, what is done with it? How is it converted back to sound waves?

A varying voltage through a coil can create a fluctuating magnetic field. This is the principle of the record head. If we apply a fluctuating magnetic field across a coil, a varying voltage will be induced in the coil. This is the *playback head*, which does just the opposite of the record head. The alignment of the magnetic particles on the tape is sensed and converted into a proportionate varying voltage that is then amplified and converted back into sound waves via a loudspeaker.

The same device can theoretically act as either a record head or a playback head, depending on whether a voltage or magnetic field is applied. In many low-priced tape recorders, the record and playback heads are combined in a single unit. This is inexpensive, but not very desirable for any serious recording work. The specific design requirements for best performance are different for record heads and playback heads. The gap size is one of the crucial factors. In addition, a combination record/playback head limits the editing possibilities.

Erase Head

There is a third type of head that is included in all practical tape recorders. This the erase head, which is vaguely similar to a record head. It is driven by a high-frequency signal called *bias*. This ensures that the magnetic particles are randomly aligned before recording.

If a preexisting alignment pattern (an earlier recording) is present, it might not be entirely canceled out by the new signal. On playback you will hear a muddy blend of the recordings, which is obviously undesirable.

Arrangement of the Heads

All tape recorders have two or three heads. Inexpensive models have two, as shown in Fig. 2-5:

☐ Erase
☐ Record/playback

Better tape recorders designed for serious use generally have three heads, as shown in Fig. 2-6:

☐ Erase
☐ Record
☐ Playback

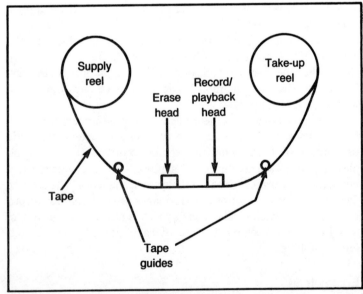

Fig. 2-5. Inexpensive recorders have two heads.

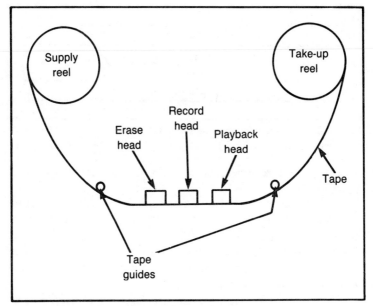

Fig. 2-6. Better tape recorders have three heads.

The heads are always positioned in the order shown in the diagrams. They cannot be arbitrarily arranged.

The erase head always comes first. It makes no sense to erase a signal on the tape after it passes the record head—you don't want to erase everything as soon as it is recorded.

When separate record and playback heads are used, the playback head is positioned after the record head. This allows you to monitor the recorded signal an instant after it has been recorded—an extremely useful feature.

A three-headed machine is not essential for creative recording, but it is a *big* help. It is not impossible for the creative recordist to do without it, but it is difficult.

TAPE TRACKS

The simplest tape recorders record a single track of information across the entire width of the tape (Fig. 2-7). This is called *single-track monaural recording*. The tape can be used more efficiently than this, however, and can hold twice as much information by dividing its width into two tracks (Fig. 2-8). First track A is recorded (or played back), then the direction of the tape is reversed by turning it over, and track B is recorded (or played back). This is called *two-*

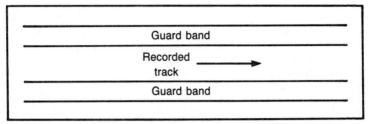

Fig. 2-7. The simplest tape recorders record a single track across the width of the tape.

track monaural recording, or sometimes *bidirectional monaural recording*.

The two tracks can also be used simultaneously in the same direction (Fig. 2-9). This is done for stereophonic recording. Track A contains the right-channel information, and track B is used for the left channel. The tape is only recorded or played back in one direction, just like the single-track monaural system. This is *two-track stereophonic recording*.

Splitting the tape width into four tracks produces a *bidirectional stereophonic recording*. This is also called *quarter-track stereophonic recording*. There are two common arrangements for the four tracks. Four-track reel-to-reel recorders interleave the tracks (Fig. 2-10). This offers maximum stereo separation. Cassette recorders, however, use adjacent tracks for each direction (Fig. 2-11). Stereo tapes may then be played on monaural cassette recorders. The four tracks may also be used in a single direction for *quadrophonic (four-channel) recording*.

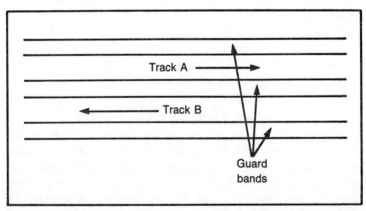

Fig. 2-8. Twice as much information can be recorded on the tape by splitting its width into two tracks.

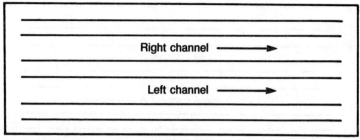

Fig. 2-9. Using both tracks at the same time allows stereo recordings.

Four tracks are about the maximum number that can be squeezed onto a standard 1/4-inch tape. (Cassettes use 1/8-inch tape.) Eight-track cartridge tapes do put eight tracks onto a 1/4-inch wide tape, but this is a convenience, rather than a high-fidelity medium.

Professional recording studios generally use 1/2-inch or 1-inch wide tape divided into 8 or 16 tracks. All tracks are recorded or played back in the same direction. The multiple tracks are for mixing purposes (see Chapter 7). Some studios even use 1-inch wide tape with up to 32 tracks.

FACTORS DETERMINING FREQUENCY RESPONSE

In any sound reproduction system, frequency response is a very important consideration. Ideally all frequencies, or at least all fre-

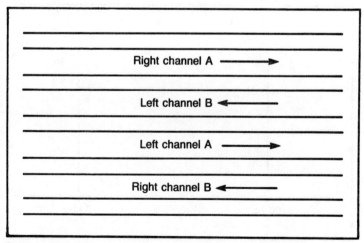

Fig. 2-10. Four-track reel-to-reel recorders interleave two pairs of stereo tracks in both directions.

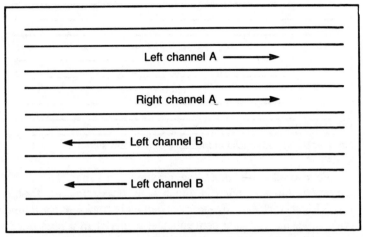

Fig. 2-11. Stereo cassette recorders use adjacent tracks for stereo.

quencies within the audible band, should be treated equally; that is, no frequency or group of frequencies should receive any more amplification or attenuation than any other frequency. This is important if the playback signal is to duplicate the original signal that was recorded.

Graphing the frequency response of an ideal recording device produces a straight line (Fig. 2-12). Unfortunately, the ideal frequency response is very easy to describe, but impossible to achieve.

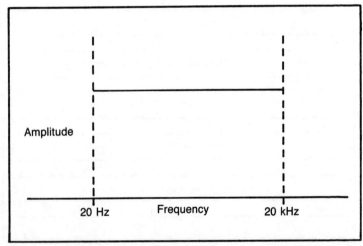

Fig. 2-12. An ideal recording device would have a perfectly flat frequency response.

Very low frequencies tend to be attenuated because the fluctuations are too slow to be decoded properly. Very low frequency components run the risk of looking like random noise to the tape recorder's heads.

More crucial is the high-frequency roll-off. The high frequencies tend to be attenuated because the fluctuations are too brief. Not enough tape passes by the record head to have room to record the cycle pattern. The magnetic particles appear to be randomly aligned, rather than aligned to a repeating pattern. High frequencies are lost in the random white noise hiss inherent in all tape recordings.

The maximum high frequency is dependent on many factors within the system. One factor is the gap size of the recording head. Narrower gaps can record higher frequencies.

Another important factor in defining the high-frequency response is the speed of tape motion. The faster the tape moves past the head, the more space there will be to encode the waveform.

Tape Speed

Home recorders generally have two speeds. These usually are 3 3/4 ips (inches per second) and 7 1/2 ips (inches per second). The slower speed is used when tape economy is the primary consideration. A recording of a given length takes twice as much tape at 7 1/2 ips than at 3 3/4 ips, but the higher speed is preferred for high-fidelity recording.

Some recorders also include a still slower speed—1 7/8 ips. This speed should be used for voice recordings only. Its frequency response characteristics are simply not suitable for music reproduction.

Professional recorders operate at higher speeds for still higher fidelity. Commonly used speeds are 15 ips and 30 ips.

The amount of tape to hold one cycle can be calculated with this formula:

$$L = S/F$$

Where L is the wavelength space on the tape, S is the tape speed, and F is the frequency of the signal to be recorded. Clearly, the greater the space for the cycle to be recorded, the more detail there can be in representing the waveshape.

Consider some practical examples. Say you want to record a 12 kHz (12,000 Hz) signal. At 1 7/8 (1.875) ips the amount of tape to record one cycle of a 12 kHz signal works out to:

$$L = 1.875/12000 = 0.000156 \text{ inch} = 0.156 \text{ millinch}$$

That isn't much space to define the waveshape.

Increasing the speed to 3 3/4 (3.75 ips) increases the tape length to:

$$L = 3.75/12000 = 0.000312 \text{ inch} = 0.312 \text{ millinch}$$

At 7 1/2 (7.5) ips:

$$L = 7.5/12000 = 0.000625 \text{ inch} = 0.625 \text{ millinch}$$

At 15 ips:

$$L = 15/12000 = 0.00125 \text{ inch} = 1.25 \text{ millinch}$$

And at 30 ips:

$$L = 30/12000 = 0.0025 \text{ inch} = 2.5 \text{ millinch}$$

This is a considerable improvement over the 0.000156-inch space allowed by a speed of 1 7/8 ips. Higher speeds allow more detail (better fidelity) for higher frequency signals.

Voltage Signal

Playback is accomplished by inducing a voltage in the playback head. This induced voltage is extremely small and must be considerably amplified to be a useful signal. Depending on the specifics of the design, the required amplification may be from 10 to 10,000. Narrow track widths and small head gaps generally call for greater amplification.

The weakness of the induced voltage signal aggravates the ever-present noise problem. In an electronic circuit, some random noise signal is generated. This noise signal is typically rather small, but when the desired signal is also very small (as in the case of the induced voltage in a playback head), the noise signal becomes significant.

To make matters even worse, the induced voltage is at its lowest for very low and very high frequencies. At such extreme frequencies, noise tends to be at its strongest.

Careful design is a must to reduce the noise signal to a tolerable level. Most modern tape recorders do quite well in this respect, but some designs do better than others.

The magnetic "charge" or flux on the tape varies with the recorded frequency. Higher-frequency signals tend to result in lower flux. This is illustrated in the graph of Fig. 2-13.

The voltage signal induced into the playback head is dependent primarily on two factors—the amount of tape flux and the frequency of the recorded signal. The playback head's response partially compensates for the decreasing flux. A typical voltage vs. frequency graph is shown in Fig. 2-14.

Equalization

To further improve the performance of the system, and to come closer to the ideal flat frequency response, some sort of equalization is usually applied to the signal. *Equalization* is a deliberate unequal amplification of different frequencies. Low and high frequencies are given more amplification, or boost, than midrange frequencies. This helps flatten out the system's frequency response (Fig. 2-15).

Unfortunately, any noise signal is equalized right along with the desired signal. As a result, low-frequency and high-frequency

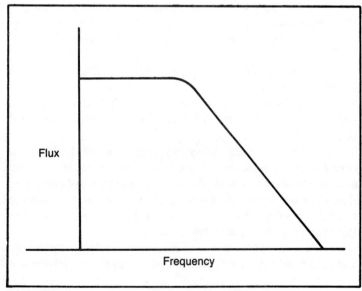

Fig. 2-13. Higher frequencies result in lower magnetic flux.

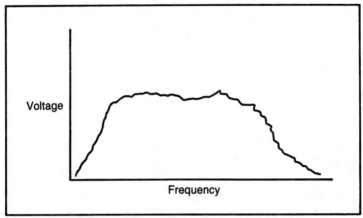

Fig. 2-14. A typical playback head's voltage vs. frequency graph.

noise is boosted (Fig. 2-16). This tends to emphasize such noise problems as amplifier hum and tape hiss.

In an attempt to compensate for this problem, most modern tape recorders apply some sort of equalization to the original signal before it is recorded. High frequencies and low frequencies are boosted. This allows the use of less equalization in the playback circuit, reducing noise emphasis.

Great care must be taken in the prerecord equalization process. If the high and low frequencies are amplified too much, the tape and/or the circuitry may be overloaded, resulting in distortion of the signal.

The audio industry has set up defined equalization standards based on what is known about sound level vs. frequency in speech and music. These equalization techniques help flatten out the system's frequency response. They also help to improve the signal-to-noise ratio.

The signal-to-noise ratio is generally given in dB (decibels). The reference is the noise level. For example, for a tape recorder with a signal-to-noise ratio of 44 dB, the recorded signal is 44 dB higher than the noise signal. If all other factors are equal, the signal-to-noise ratio is degraded at least 3 dB if the track width is halved (four tracks as opposed to two tracks).

While there is considerable variation due to design factors, the signal-to-noise ratio for typical sound reproduction systems are:

- ☐ 78 rpm phonograph records   35-40 dB
- ☐ Inexpensive tape recorders   35-45 dB

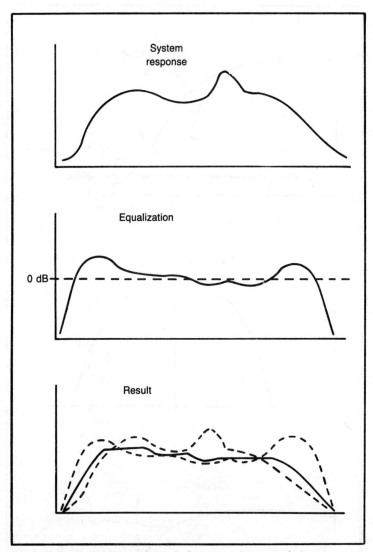

Fig. 2-15. Equilization is used to help flatten out the system's frequency response.

- [ ] Quality home tape recorders — 45-55 dB
- [ ] Modern LP (33 1/3 rpm) phonograph records — 50-65 dB
- [ ] Studio tape recorders — 50-75 dB
- [ ] Digital studio recorders — 85-95 dB

A good equalization system can give very close to flat frequency

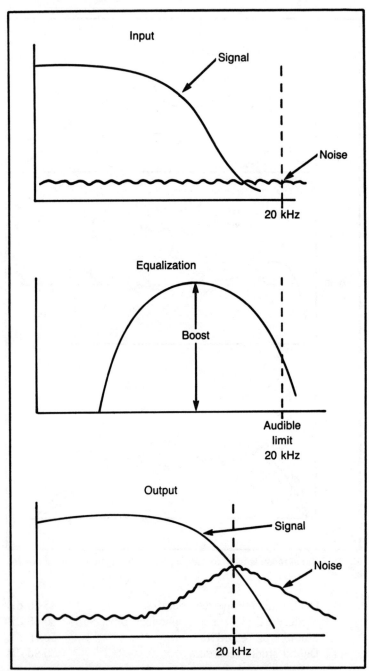

Fig. 2-16. Straight equalization boosting can also boost noise.

response; however, a truly flat response is not possible. There are simply too many factors to account for. The frequency response of a tape recorder is usually given in this form:

$$X \text{ to } Y \text{ Hz } \pm - Z \text{ dB}$$

where X is the lowest frequency, Y is the highest frequency, and Z is the maximum variation from flat response in the specified frequency range. For example, a typical tape machine might have a specified frequency response of 60 to 15,000 Hz ± − 3 dB. This means that any frequency in the 60 to 15,000 Hz range will be no more than 3 dB lower or 3 dB higher than the nominal flat frequency response level. A difference of 3 dB is just barely noticeable.

In shopping for a tape recorder or other sound equipment, look for the widest possible specified frequency range with the lowest dB variation figure. As a rule of thumb, consider the following dB values:

- ☐ 1 dB—excellent response; just barely noticeable by an expert under ideal listening conditions
- ☐ 3 dB—good response; noticeable under normal listening conditions
- ☐ 6 dB—poor response; an unmistakable change in sound level

Frequencies outside the specified range can be reproduced, but the variation in level may be greater than specified. For instance, in our sample machine with a frequency response specification of 60 to 15,000 Hz ± − 3 dB, a 17,500 Hz signal might be reproduced at a level that is 5 dB lower than the nominal flat response level.

Few home tape recorders are specified for use above about 15,000 to 16,000 Hz. This might seem to be a major disadvantage, since the human ear nominally can hear frequencies up to 20,000 Hz. Fortunately, this is not as big a problem as it might seem at first. The maximum audible frequency tends to decrease with the listener's age, and few serious listeners can really hear much above 15 to 16 kHz. Besides, few musical sounds contain strong frequency components above about 10,000 Hz or so. A trained listener will be able to perceive some difference, but for general listening, a frequency response of 60 to 15,000 Hz is entirely acceptable.

RECORD BIAS

An extremely important factor in reproducing sound from a tape recorder is the *record bias*. This is an ultrasonic (inaudible) high-frequency signal, generated by an internal oscillator, that is mixed with the signal to be recorded at the record head. Without getting into the theoretical aspects, which are somewhat complex, the record bias increases the high-frequency response and decreases distortion.

The exact frequency of the record bias signal is not particularly crucial. It must only be high enough not to cause audible beating effects. Beating occurs at the difference between two frequencies. For example, say you are recording a 12 kHz signal, and the record bias is set at 17 kHz. A spurious beat frequency of 5 kHz might appear in the recorded signal. Clearly this is extremely desirable.

In most modern tape recorders the record bias frequency is about 70 kHz to 150 kHz. This range of frequencies does the job well, without obtrusive side effects. Often the same high-frequency signal that is used by the erase head to erase a previously recorded signal is used as the record bias signal.

While the record bias frequency is not crucial, the amplitude of the record bias signal is. If the record bias level is too low, or if there is no bias used at all, the recorded signal will be distorted. This distortion may be fairly mild, resulting in a rather bland sound, or it may be quite severe. In addition, the signal-to-noise ratio will be significantly degraded if the bias is not strong enough. On the other hand, if the bias signal level is too high, the record head tends to function as an erase head, especially at high frequencies. The signal will be erased as it is being recorded.

Unfortunately, determining the best record bias level is not an easy task. Many factors must be considered. These include:

- ☐ Design of the record head
- ☐ Design of the recorder's electronic circuitry
- ☐ Characteristics of the tape being used
- ☐ Frequency of the signal being recorded

The first two factors remain fairly constant, although some of these characteristics might change as the components age. Tape characteristics can differ significantly from brand to brand, however, or even from batch to batch for the same brand. The frequency of the signal being recorded will vary a great deal. Because of these problems, the theoretical "ideal" record bias level is an "impossible dream." A good compromise can be obtained without too much trouble. As a rule, the record bias signal's amplitude is approximately 10 times the amplitude of the signal to be recorded.

TAPE FORMATS

Several basic configurations are used for modern tape recorders. The differences between them are essentially convenience, at least theoretically. Most popular tape recorders today are one of three types:

- ☐ Reel-to-reel
- ☐ Cassette
- ☐ 8-track cartridge

Reel-to-Reel Tape Recorders

The most basic configuration is the reel-to-reel approach (Fig. 2-17). The tape is wound onto a plastic or metal reel. When loaded on the machine, the tape passes through several guides, the heads, and a capstan/pinch roller assembly. The capstan and pinch roller pull the tape across the heads at the desired uniform speed. The tape is then wound onto a second plastic or metal reel called the take-up reel.

Once the entire tape is wound onto the take-up reel, it may be rewound back onto the original reel to be played again at a later date. Some purists suggest that the tape be stored on the take-up reel and rewound just prior to playback. This is because the high-speed modes (rewind and fast forward) wind the tape less evenly, increasing the chance of print-through during storage. (Print-through and related problems are discussed in Chapter 7.) Other experts feel that it really doesn't make much difference.

For bidirectional recorders, where the tape is recorded with tracks running in both directions, the rewinding question becomes moot. Once the tape is wound off entirely on the take-up reel, the two reels reverse their positions. The now full take-up reel becomes the new source reel, and the now empty original source reel becomes the new take-up reel. The second side of the tape can now be played in the same manner as the first.

Some bidirectional tape recorders include an automatic reverse feature. A small metallic tab is pasted on the end of the tape. When this metallic tab is sensed, the recorder's motors reverse direction, and the tape starts running through the machine backwards to play the other side. The complex mechanical switching system required

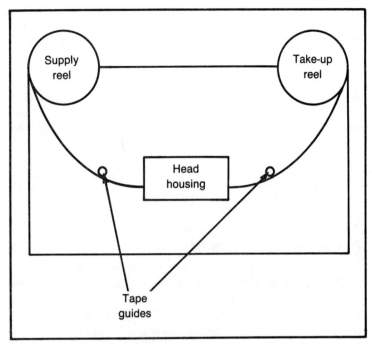

Fig. 2-17. The most basic configuration is the reel-to-reel.

for this feature makes it a fairly expensive option.

A typical home reel-to-reel tape recorder is shown in Fig. 2-18. The type of tape used on this sort of machine is shown in Fig. 2-19. Figure 2-20 shows a professional reel-to-reel tape recorder.

Reel-to-reel recorders have many advantages. They offer the best sound quality on a dollar-for-dollar basis. Editing is relatively easy. The simple drive system minimizes jamming problems. The heads are usually easily accessible for cleaning and other maintenance. On most reel-to-reel machines, two or three speeds are offered.

There are also some disadvantages. Reel tapes are relatively large and bulky. The tape is exposed and can therefore be contaminated by dust or oils from your fingertips. The tape must be manually threaded, which can be a minor nusiance. Not many prerecorded tapes are available in this format. Low-end (inexpensive) reel-to-reel recorders are no longer manufactured.

Cassettes

Cassette tapes are by far the most popular format today. A typi-

Fig. 2-18. A typical reel-to-reel tape recorder.

cal cassette tape is shown in Fig. 2-21 and a cassette tape recorder in Fig. 2-22. A cassette is basically a miniature reel-to-reel system housed in a plastic case. The internal structure of a cassette tape is illustrated in Fig. 2-23.

Cassette tape is approximately 1/8-inch wide. It is divided into two (monaural) or four (stereo) tracks (Fig. 2-24). The arrangement of the tracks allow full compatibility between monaural and stereophonic machines. This was a requirement of the original patent holder, Phillips.

Originally designed for voice-only "note-taking," the convenience of the tape cassette quickly caught on. Improved designs

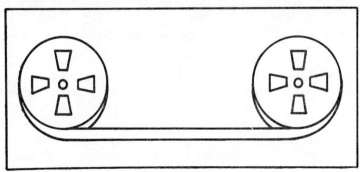

Fig. 2-19. Reel-to-reel tape is wound on an open plastic or metal reel.

Fig. 2-20. Professional studios use larger, more advanced reel-to-reel tape recorders.

made it more suitable for high-fidelity music recording. Special tape formations and noise reduction circuits made the cassette's sound quality more or less comparable to that of reel-to-reel recorders.

Small cutouts in the back of the cassette housing are used for automatic sensing by the tape machine. For example, when the RECORD button is pushed, a small finger checks for a plastic tab in the back of the housing. If the tab is present, the machine goes

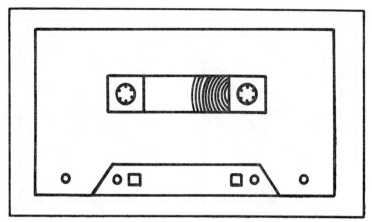

Fig. 2-21. This is a typical cassette tape.

into the record mode. If the tab has been broken off, the record mode can not be activated, protecting the tape from accidental erasure. If the tab is accidentally broken off, it can be simulated by placing a strip of cellophane tape over the opening.

A second sensor found on many better cassette recorders checks for another tab. If this tab is missing, the recorder assumes the tape is chromium dioxide and automatically adjusts its equalization accordingly.

Fig. 2-22. A typical cassette tape deck.

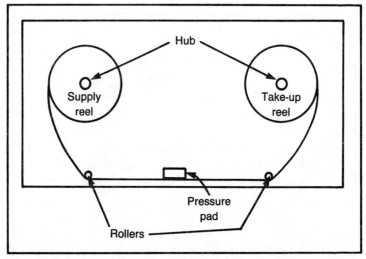

Fig. 2-23. A cassette is rather like a miniature reel-to-reel system in a plastic housing.

The advantages of cassettes include their small size, allowing for convenient storage and transportation. A wide variety of cassette recorders are available, including inexpensive portable units, car stereos, and home stereo decks. Tapes are fairly inexpensive and easy to use. The actual tape is never handled by the user. Hundreds of prerecorded programs are available in this format.

Cassettes do have their share of shortcomings, however. Their narrow width and slow speed (1 7/8 ips) demand a lot of extra cir-

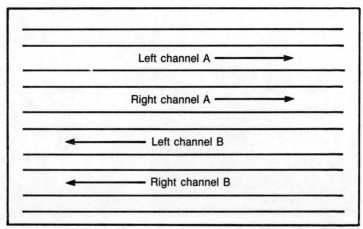

Fig. 2-24. Cassette tape is divided into two monaural or four stereo tracks.

cuitry to produce acceptable sound quality. This increases equipment costs and results in more stages that could develop defects. They are also next to impossible to edit. Finally, the enclosed housing is prone to jamming.

8-Track Cartridges

A third popular tape format is the 8-track cartridge (Fig. 2-5). This type of tape is contained in a plastic housing in an endless loop arrangement. There is no actual beginning or ending to the tape. It can be played continuously over and over.

The tape is divided into eight tracks, representing four stereo pairs (Fig. 2-26). A metallic tab is placed across the tape at the end of the loop. When this tab is sensed, the recorder physically moves the head(s) to play a new pair of tracks.

Eight-track recorders are not as popular today as they were a few years ago. Cassettes have been taking over the low-end market this tape format once appealed to. About the only significant advantage of the 8-track cartridge is that the endless loop approach

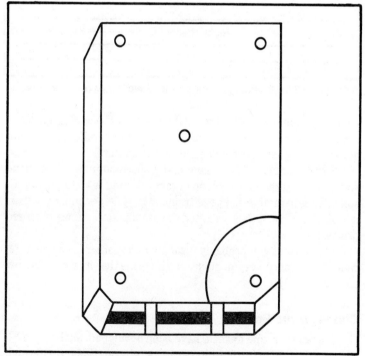

Fig. 2-25. A third popular tape format is the 8-track cartridge.

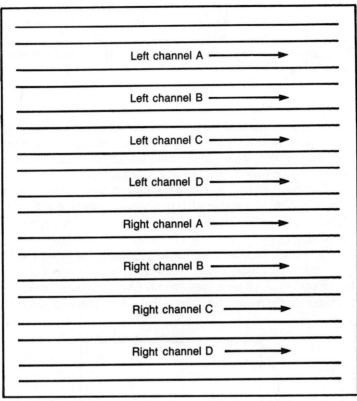

Fig. 2-26. An 8-track cartridge's tape is divided into four pairs of stereo tracks.

allows for continuous "background music" playback. The enclosed housing also makes the tape easy to store and handle, but it is not quite as compact and convenient as the cassette.

Eight-track cartridges have many shortcomings. There is no way to rewind the endless loop of tape. Editing is virtually impossible. Tape jamming happens frequently. The declining popularity of this format has resulted in a reduction of prerecorded programs.

This format will probably disappear altogether in the next decade. It was once very popular, but it seems that its time has now passed.

Choosing the Format

A few other tape formats have been developed, such as the Elcassette, but none of these have yet caught on. Since the 8-track cartridge is apparently going the way of the 78 rpm disc, the choice is really between the reel-to-reel or cassette.

If you are mainly interested in dubbing record albums or playing prerecorded tapes, cassettes are probably your best choice just because of their convenience. For creative recording their small size and enclosed housings become serious liabilities, however. They are extremely difficult to edit and creatively manipulate. There is still a lot of creative recording you can do with a cassette machine that will be covered in this book. For serious recording work, however, the old-fashioned reel-to-reel recorder still wins hands down.

Chapter 3

DB and VU

This chapter examines two important sound/signal measurement systems that are widely used in recording work. The first is the dB (decibel) and the second, the VU (volume units) measurement. The chapter also looks at signal-to-noise (S/N) ratios.

DECIBELS

Decibels are a comparative measurement. Unless the reference level is specified, any dB value is meaningless. Before making any use of the decibel system, the 0 dB (reference) level must be defined. Then, if a second signal has a dB greater than 0 (positive), it is higher than the reference signal. If the second signal's dB value is less than 0 (negative), the level is lower than that of the reference signal.

In some applications a standard 0 dB reference level can be assumed. For SPL (sound pressure level) readings, for example, 0 dB represents the theoretical threshold of human hearing. The *threshold of human hearing* is the lowest level the average person can hear. Though some people can't hear anything at that weak a level, others can hear even softer sounds. The standard is simply a convenient average. It represents the equivalent of 0.000000000001 watt.

Using this standard reference, various typical sounds can be compared in a useful fashion:

0 dB	threshold of hearing
10 dB	quiet recording studio
20 dB	quiet living room
30 dB	quiet office
40 dB	subdued conversation
50 dB	average office
60 dB	average conversation
65 dB	small orchestra
75 dB	average factory/busy street
85 dB	heavy truck traffic
95 dB	subway
110 dB	power tools
115 dB	thunder
120 dB	airport runway
125 dB	pneumatic hammer
135 dB	threshold of pain
155 dB	jet engine (close)

The ear can detect a very large range of amplitudes. (Remember, 6 dB is a doubling of volume.) Long term exposure to sounds above about 100 dB can result in permanent damage to the ear. Sounds above the threshold of pain (approximately 135 dB) should be avoided, since even short-term exposure can be damaging.

It is very important to remember that the 0 dB point is arbitrarily selected. Other 0 dB references may also be selected for other applications. The 0 dB level on most tape recorder meters has absolutely no relation to the 0 dB of the SPL scale just described.

Another frequently used scale is often called the *dBm scale*. This scale is used on many meters in audio work. In this scale, power levels are referenced to 1 milliwatt across a 600 ohm line. (600 ohms is the standard line impedance for most—but not all—audio equipment.)

The difference between two power (or sound) levels in dB can be calculated with this formula:

$$10 \text{Log} (P1/P2)$$

where P1 and P2 are the power levels being compared. The logarithm (to the base of 10) is taken of the fraction ratio, then the result is multiplied by 10 to get the difference in dB.

VU METERS

Most tape recorders feature signal level meters that are calibrated in arbitrary volume units, or VUs. This provides a convenient method of monitoring signal levels while recording. (Most VU meters don't give very meaningful readings during playback, when they generally aren't needed anyway.)

If the signal being monitored is a steady tone, a VU meter will respond pretty much like a dBm meter. Real musical signals are much more complex, however, and filled with rapid changes. Most meter movements cannot respond fast enough to indicate brief peaks in the music signal. By the time the meter's pointer starts to move up the scale in response to the peak, the peak is over.

VU meters are designed with special ballistic characteristics that allow the pointer to respond to music signals in a manner similar to the human ear. Consequently a VU meter can give a more meaningful indication than a basic dBm meter for music signals.

The range of a VU meter varies from manufacturer to manufacturer. Typically VU meters read from -20 to $+3$ VU. The 0 VU point is defined as the maximum signal level that can be recorded without an unacceptable amount of distortion. This is a fairly arbitrary value. For one thing, just what is an acceptable amount of distortion? 1 percent? 3 percent? In addition, the amount of distortion for a given recording level can vary a great deal with different types of tape.

The 0 VU point is an arbitrary compromise, but it's a good reference for monitoring the signal being recorded. Usually 3 percent distortion is considered the cutoff point of acceptability.

The standard VU meter is designed with an input impedance of 3900 ohms. It is set up to be used with a series resistor with a value of 3600 ohms.

Most VU meters do not respond fully to very brief peaks. The instantaneous signal may be 10 to 20 dB higher than what is indicated by the meter.

Some more deluxe recorders feature so-called "peak-reading" meters. These meters are designed to have a very rapid rise time (to respond to a sudden peak) and a slow decay time (to hold the reading long enough to be read). Even these meters aren't completely responsive to brief peaks, but they do a very good job overall.

The purpose of the VU meter is to set the signal level fed to the recording head to maximize the signal-to-noise ratio (by recording as strong a signal as possible) while minimizing distortion (by not recording too strong a signal). The distortion comes from clipping in the electronics circuitry and/or saturation of the tape.

S/N SPECIFICATIONS

One of the most important specifications for a tape recorder (or any audio equipment) is the S/N (signal-to-noise) ratio. This is a dB value defining how much higher the signal can be compared to the noise generated by the system without unacceptable distortion—again a compromise because it is impossible to concretely define how much distortion is too much.

Be careful when reading manufacturer's specification sheets. The S/N specification is only meaningful when you know the distortion level at the 0 VU point. If manufacturer A sets the 0 VU point for 1 percent distortion, and manufacturer B uses 3 percent distortion as the standard, manufacturer B's S/N ratio will look better than manufacturer A's (all other factors being equal). B's measurements will have an extra 6 to 9 dB just because of the difference in the definition of "acceptable" distortion.

Another potential problem in S/N measurements comes from the way the noise level is measured. In absolute terms, low-frequency noise (primarily from the 60 Hz line current and its low harmonics) is considerably stronger than the noise in the high-frequency range (hiss). A simple direct noise measurement is dominated by the low-frequency noise content. The human ear tends to be much more sensitive to high-frequency noise, however, so the low-frequency noise is relatively insignificant.

Let's assume that recorder A and recorder B have the same amount of 60 Hz hum, but A has about 5 dB more hiss than B. Simple direct noise readings indicate that the two recorders have about the same amount of noise, but B sounds noticeably better than A.

Most manufacturers use a weighting curve (high-pass filtering) when making noise measurements to get a value that more closely approximates the perceived amount of noise. Unfortunately, several different weighting curves are used, making it difficult to compare specifications from different sources.

The S/N ratio also varies with the signal being recorded. The differences between steady tones and various types of music can be considerable.

Despite the importance of the S/N ratio in defining recorder performance, the lack of standardization makes it a very imprecise specification. Take S/N specs with a large grain of salt unless the specifics of the measurement are defined.

Chapter 4

Microphones

In live recording a device is needed to convert the sound signals into electrical signals that can be recorded. This device is a *microphone*.

In schematic diagrams, the symbol shown in Fig. 4-1 is generally used to represent a microphone. The word microphone is often shortened to "mike." There are many different types of microphones available, each with their own individual characteristics.

CARBON MICROPHONES

Perhaps the simplest type of microphone is the carbon microphone. The basic construction of this device is illustrated in Fig. 4-2.

Essentially, a carbon microphone consists of a small container with carbon discs on either end and is filled with tiny carbon granules. One of the end discs is rigidly held in a fixed position, while the other is movable. The movable disc is connected to a flexible diaphragm.

Fluctuations in air pressure (sound) move the diaphragm, and therefore the second carbon disc, back and forth. This action puts greater or less pressure on the carbon granules within the container. The changes in the density of these packed particles change the effective resistance.

If this assembly is placed in series with a dc voltage source,

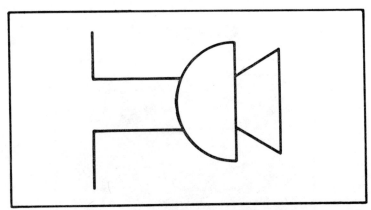

Fig. 4-1. This symbol is often used to represent a microphone in schematic diagrams.

such as the small battery in Fig. 4-3, the voltage drop across the microphone will vary along with the changes in the resistance of the carbon particles, which varies with the sound pressure at the diaphragm. As a result, the output voltage varies in step with the sound waves reaching the microphone element. This is the electrical equivalent of the acoustic energy.

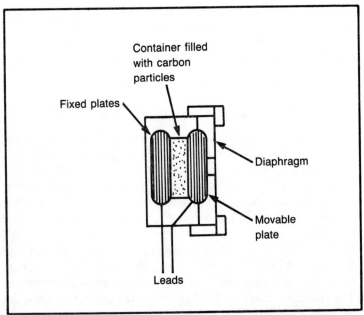

Fig. 4-2. The simplest type of microphone is the carbon microphone.

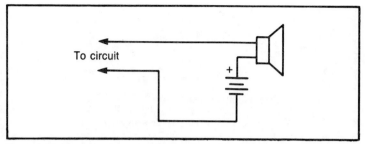

Fig. 4-3. The carbon element acts like a variable resistance in series with a voltage source.

The primary advantages of carbon microphones are their relatively low cost and the high signal level output they provide. A carbon microphone can produce a stronger electrical signal than any other microphone available. These advantages don't make carbon microphones the best choice, however. Not by a long shot. They have a number of significant disadvantages too.

Carbon microphones require an external voltage source, which can be a nuisance in many systems. Many signal inputs are not designed to handle the high signal levels produced by carbon microphones. More important limitations involve the carbon microphone's narrow frequency response and rather high noise and distortion levels.

Carbon microphones are only suitable for noncrucial voice-only applications. They are commonly employed in telephone handsets.

CRYSTAL MICROPHONES

Another popular low-cost microphone is the crystal microphone, which depends upon the piezoelectric effect for its operation. The basic construction of a crystal microphone is illustrated in Fig. 4-4.

The sound pressure on the diaphragm creates a mechanical stress on the crystal element. The *piezoelectric effect* means that a mechanical stress placed across one axis of a crystal produces a comparable electrical stress along a perpendicular axis. Thus, a voltage is generated that varies in step with the mechanical stress (motion of the diaphragm in response to sound pressure).

While resonant crystals used in many high-frequency circuits are generally made of quartz, the crystal elements used in this type of microphone are usually made of Rochelle salt.

Unlike the carbon microphone, the crystal microphone requires no external voltage source. The frequency response of a typical

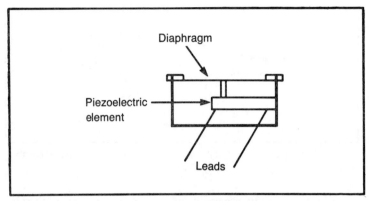

Fig. 4-4. A crystal microphone is based on the piezoelectric effect.

crystal microphone is fair, but it is not really good enough for high-fidelity work. The output signal level is still fairly high for this type of microphone.

One of the biggest disadvantages of the crystal microphone, aside from its somewhat limited frequency response, is that it is rather fragile and easily damaged. Also, the Rochelle salt crystal can absorb moisture, which would ruin the microphone. These two problems can be alleviated by replacing the crystal with a somewhat more rugged ceramic element. This makes it a *ceramic microphone*. Except for the difference in the element material, crystal and ceramic microphones function in the same way. Both are widely used in communications (radio) applications.

DYNAMIC MICROPHONES

Probably the most popular type of general purpose microphone is the dynamic microphone. The basic structure of this device is illustrated in Fig. 4-5.

In this type of microphone, the diaphragm is connected to a small coil (inductor). This coil is suspended so that both it and the diaphragm can move freely in response to the sound pressure. The coil is moved back and forth within the magnetic field of a permanent magnet. This induces a voltage in the coil that varies in step with its movement that is, in turn, defined by the sound pressure striking the diaphragm.

Basically, the dynamic microphone functions as a loudspeaker in reverse. Even a small speaker might be used as a low-grade dynamic microphone.

The output signal level from a dynamic microphone is typically

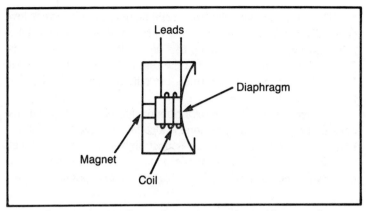

Fig. 4-5. The dynamic microphone is the most popular type for general-purpose use.

rather low, but the frequency response is quite good. This type of microphone is also fairly sturdy and durable.

RIBBON MICROPHONES

Closely related to the dynamic microphone is the ribbon microphone. The basic construction of this type of microphone is shown in Fig. 4-6. It has a corrugated aluminum ribbon that is moved by the sound pressure within the magnetic field of a permanent magnet. A small ac voltage is induced in the ribbon by this action.

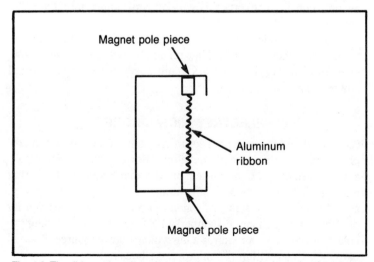

Fig. 4-6. The ribbon microphone is a variation on the basic dynamic microphone.

The output signal level from a ribbon microphone is very low. Generally a step-up transformer or a preamplifier must be used to raise the signal's amplitude up to a usable level. A step-up transformer is often contained within the case of the microphone itself. Even with a step-up transformer, the output signal level is fairly low, but usable.

The frequency response of a ribbon microphone is typically excellent. This type of microphone is also fairly rugged, but the ribbon element must be protected against high sound pressure levels or blasts of wind. This type of microphone is employed only in studio environments, although there are some exceptions.

CONDENSER MICROPHONES

A condenser microphone is based on the behavior of capacitors (once known as "condenser"). A *capacitor* is simply two metallic plates separated by an insulator (dielectric). Ac signals can pass through a capacitor, but dc signals are blocked.

The basic construction of a condenser microphone is illustrated in Fig. 4-7. Two small metallic plates are slightly separated by air or some other flexible insulator (dielectric). One of the plates is rigid and fixed in position. The other plate is flexible and functions as the microphone's diaphragm. As sound pressure forces the diaphragm to move back and forth, the distance between the plates varies. This alters the capacitance. A small circuit within the microphone's housing converts this varying capacitance into a varying voltage.

Condensor microphones offer very low distortion and an excellent frequency response. They tend to be rather expensive, however. In addition, the internal circuitry requires its own external power source.

ELECTRET MICROPHONE

Closely related to the condensor microphone is the electret microphone. In this fairly recently developed device, the polarizing voltage is permanently impressed on the diaphragm plate during manufacture. No external high-voltage power supply is required, as is the case for the standard condenser microphone. The majority of electret microphones do contain a FET impedance matching circuit, however, which requires a low-voltage power source. A small battery is used within the microphone housing.

Electret microphones are increasing in popularity in semiprofes-

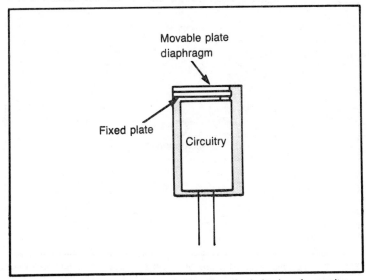

Fig. 4-7. A condenser microphone is based on the behavior of capacitors.

sional and amateur applications. They are even used in some professional recording studios. They offer excellent frequency response, thanks to the low mass of the diaphragm. In addition, this type of microphone can provide a very clean sound. It can clearly respond to fast, transient sounds.

MICROPHONE PICKUP PATTERNS

Microphones can be classified according to which method they use. The microphones described in the preceding sections are the most commonly used, although there are others. Another important way to discriminate between microphones is the pickup pattern.

Some microphones can pick up all sounds equally from all directions. This type of microphone is said to have an *omnidirectional* pickup pattern. The omnidirectional pickup pattern is illustrated in Fig. 4-8. It doesn't matter where the sound source is in relation to the microphone. A sound source behind the microphone will be picked up as well as a sound source the same distance in front of the microphone.

While useful in some applications, an omnidirectional microphone's pickup pattern can be a major disadvantage if you want to pick up one sound source out of many. In this case use a *unidirectional* microphone. This pickup pattern is illustrated in Fig. 4-9. A unidirectional microphone picks up sounds from only one direc-

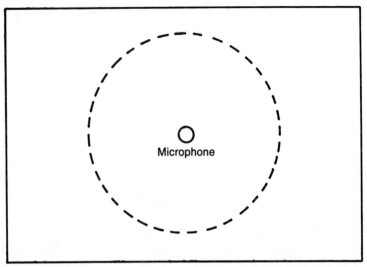

Fig. 4-8. An omnidirectional microphone picks up sounds equally well in all directions.

tion. A sound source behind the microphone is more or less ignored. A unidirectional microphone can't totally block out sound sources outside its nominal pickup pattern; some room echoes will reach the microphone from the front. This concept is illustrated in Fig. 4-10. Nevertheless, sounds from outside the microphone's defined

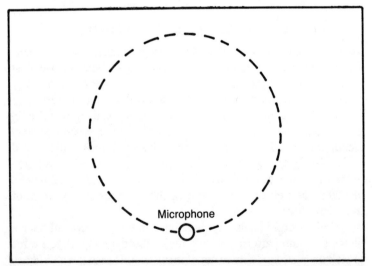

Fig. 4-9. A unidirectional microphone picks up sounds from the front better than sounds coming from other directions.

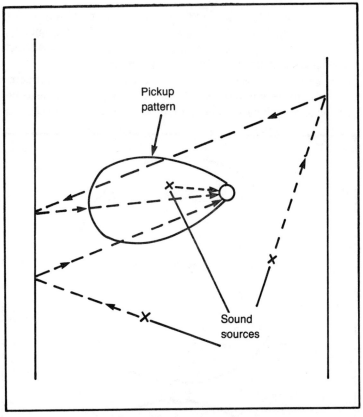

Fig. 4-10. Some sounds from behind and to the sides of a unidirectional microphone are probably picked up due to room echoes.

pickup pattern are greatly attenuated. In a sense, this type of microphone can "focus in" on a single sound source or group of sound sources.

Still other microphones feature a *bidirectional* pickup pattern (Fig. 4-11). This pickup pattern can be simulated with a pair of back-to-back unidirectional microphones. Bidirectional microphones are often used for radio performers. Several actors can share a single microphone without crowding. Some singing groups also employ bidirectional microphones for the same reason.

A variation on the unidirectional pickup pattern is the *cardioid* pickup pattern (Fig. 4-12). This type of microphone also concentrates on sound sources directly in front of it, but it is more sensitive to sound sources at the sides. The name "cardioid" comes from the heart-shaped pickup pattern.

When cardioid microphones are used, the proximity effect should be considered. Within about 2 inches of the microphone, the bass response is exaggerated, distorting the signal. This effect does not occur with omnidirectional microphones.

A speaker or singer should not be placed too close to any microphone. Sibilants and plosives (Ss and Ps for example) can be exaggerated and produce an unpleasant effect. A good windscreen can

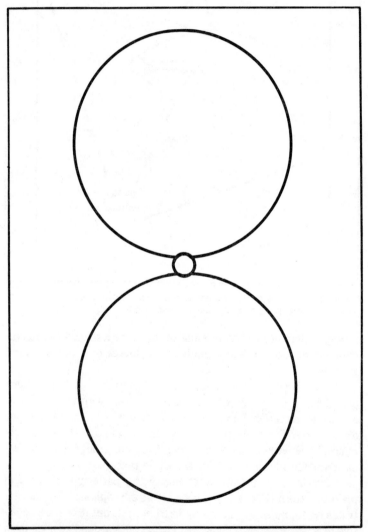

Fig. 4-11. A bidirectional microphone functions like a pair of back-to-back unidirectional microphones.

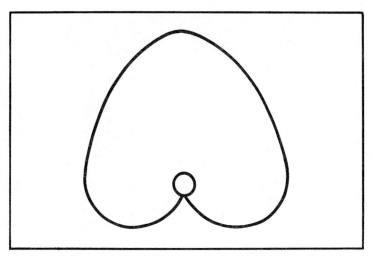

Fig. 4-12. A cardioid microphone is more or less unidirectional with a heart-shaped pickup pattern.

reduce this problem, but even better results can be achieved by using the windscreen and staying back a few inches from the microphone. Singers who appear to be "eating" their microphones are exhibiting very poor microphone technique.

Chapter 5

Setting Up a Home Recording Studio

To do any serious recording work, you need some sort of studio. There are an infinite number of possibilities, depending on your specific needs and budget. Field or live recording is covered in Chapter 6. Even if you plan to do all of your recording on location, a permanent studio/mixing room is still very important for creative recording.

Few of us can afford to sink several hundred thousand dollars into the major construction and high-grade acoustics of a professional studio. That doesn't mean you are out of luck. An adequate, semiprofessional-grade studio doesn't have to be that expensive.

This chapter describes what you can do to create a studio in your own home. How elaborate you get depends on what you can afford and how much room you can devote to your studio. Some hobbyists might have to satisfy themselves with a temporary "portable" studio that can be set up and then put away so the room can be used for other purposes.

SELECTING A TAPE MACHINE

There are a lot of choices involved in selecting a tape recorder. Many of the decisions depend more on personal preferences than anything else. Get a machine you like, whatever that may mean to you. If you don't feel comfortable with your tape recorder, it is likely to end up as an expensive dust collector or a major frustration generator.

Price is probably an important factor. If you can't afford a specific machine, it doesn't matter how good it is or how many special features it has—you can't have it if you can't afford it. Personally, I would love to have a studio-grade, 16-track recorder, but it is far out of my price range.

On the other hand, don't base your decision solely on the current condition of your bank account. Let's assume that you don't particularly like recorder A, which costs $450, and dream of owning recorder B, which has a $600 price tag. Your budget is $500. Under these circumstances, it would not be a good bargain to buy recorder A. You won't be satisfied with it and will either give up on recording or suffer until you can scrape up enough money for the machine you wanted in the first place, which will probably cost $750 by then. Then recorder A will be put on the shelf, its purchase price wasted.

If you are faced with this kind of situation, find some way to add $100 to your budget or wait until you've saved up enough to get what you need. Never buy inadequate equipment just because it's cheap. That's never a bargain.

On the other hand, you have to make compromises. If recorder B at $600 does the job, but recorder C at $900 has a lot of nifty extra features and gives a somewhat better performance, then your current economic situation may well be the deciding factor. The important thing is to make sure you get what you need. Compromise on the extras, not the essentials.

For mixing and dubbing, two inexpensive (but adequate) tape decks are preferable to one super-deluxe (and expensive) tape recorder. Two $500 decks could be worth more to you than one $1300 deck. If you can afford three decks, so much the better. You can do almost everything with two machines, but three is even more convenient. The second (and third) deck can be used primarily for straight playback (see Chapter 9), so it can be a straightforward machine without many special features. This can help cut costs significantly.

If you can only afford one tape recorder now, that's alright. There's still a lot you can do, and you can certainly add a second or even a third machine at a later date.

For creative recording, a reel-to-reel machine is best. A good cassette deck can be used, especially if you already have one as part of your stereo system, but it tends to be more difficult to work with. If you're shopping for a recorder specifically for creative or serious recording work, as opposed to making copies of record albums or playing prerecorded tapes, I'd strongly advise against a cassette deck. It is not designed for the kind of work discussed in this book.

A good reel-to-reel deck should have at least two speeds. Remember, the higher the speed, the better the high-end frequency response. Most home decks offer 7 1/2-ips and 3 3/4-ips speeds. Some also include 1 7/8-ips speeds, but that speed is virtually useless for your purposes. Better semiprofessional recorders have 7 1/2-ips and 15-ips speeds. Some machines have a manual variable speed control, which can come in handy for many special effects.

The maximum reel size is another factor to consider. Small portable recorders that take only 3 1/2-inch or 5-inch reels were once popular. Such small reels don't hold much tape. Cassette recorders have practically driven portable reel-to-reel recorders off the market. This is no great loss because they usually weren't much good.

Most home decks hold 7-inch reels, as shown in Fig. 5-1. Semiprofessional machines hold the 10 1/2-inch reels that are used in most professional studios (Fig. 5-2).

You might also consider a noise reduction circuit, such as Dolby (see Chapter 10). This might be considered dispensable for many applications. Outboard noise reduction units can be added to the system later.

Track width is yet another factor. Typically a half-track (two

Fig. 5-1. Most home tape decks hold up to 7-inch reels.

Fig. 5-2. Professional recorders can hold 10 1/2-inch reels.

channels on the 1/4-inch wide tape) recording has a S/N ratio that is about 3 dB better than a comparable recording on a quarter-track (four channels on the 1/4-inch wide tape) machine. Quarter-track recording can be more economical in terms of tape use because one stereo program can be recorded in one direction and a second recorded on the other "side" of the tape. A half-track recording can hold only one stereo (two-channel) program. There is no place to put the second side.

For any editing work, however, you can use only one side of the tape. The second side would be randomly interrupted for each of the cuts made on the first side.

The wider width of each track makes unusual splices (see Chapter 8) easier with half-track recordings. Quarter-track recorders tend to be somewhat less expensive and more readily available than half-track units. Either type may be used for creative recording work. A full-track recorder offers an even better S/N ratio, but the limitations of its monaural-only operation make such a machine impractical for creative recording. Few, if any, full-track tape recorders are being marketed today.

One type of quarter-track recorder would probably win out over a comparable half-track machine for the creative recordist. A *four-channel* recorder can record up to four discrete channels running in the same direction. The added mixing capabilities (see Chapter 9) more than outweighs the 3 dB loss in the S/N ratio, especially since an external noise reduction circuit (see Chapter 10) can easily make up the difference, and then some.

Quadrophonic (four-channel) sound never really caught on for home stereo systems. Part of the reason was that there were two major incompatible systems. Each system required its own type of encoded records. The average consumer decided to sit back and wait to see which of the two systems would eventually win out (and therefore have more records available). Because sales were so sluggish, both systems soon died out, and quadrophonic sound was soon forgotten. (This is certainly a shame. A good four-channel recording can be quite exciting.)

Four-channel recorders are still available. The demand for these machines comes mainly from creative recordists who want to do multitrack mixing. A four-channel tape recorder is by no means essential, but if all other factors are nearly equal, four-channel capability is a big plus.

Definitely get a three-head machine. I don't think two-head reel-to-reel recorders are manufactured anymore. Even if they are, such machines are not really suitable for serious recording work. You can get by with two heads, but again, you will be severely limited in your creative recording capabilities.

Your tape recorder should also have a synchronization switch. A sync switch allows you to record on one channel while listening to a second channel in synchronization. This is done by using the record head as a playback head. Why not just use the playback head? Consider the typical head arrangement, shown again in Fig. 5-1. The tape passes over the playback head a few milliseconds before it passes over the record head. If you play a second part in time with the first part while listening to it from the playback head, the result will be noticeably out of kilter (Fig. 5-3). The effect is exaggerated for slower tape speeds.

The playback quality of the sound is not as good when the record head is used. When you use the sync switch, however, you are more concerned with timing than high fidelity. Synchronization is discussed further in Chapter 9.

Most modern reel-to-reel tape recorders have three motors. These machines are more sturdy and durable than comparable two-motor machines because each motor has to do less work. If you happen to come across an old one-motor tape recorder, pass it by without a moment's hesitation (unless you're into collecting an-

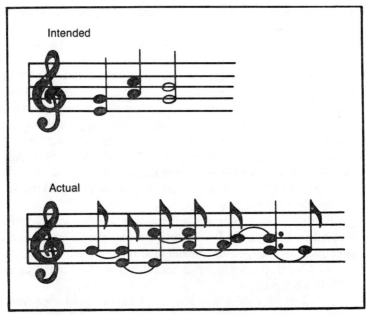

Fig. 5-3. The space between the record and playback heads creates problems for synchronization.

tiques). Even if it works, the odds are the motor will eventually burn itself out. Also, such machines are so old that getting replacement parts for repairs is nearly impossible.

Many modern tape decks feature *logic control* of their operations. This is another desirable feature linked to reliability. An ordinary drive system can be seriously damaged by misoperation—for example, going directly from rewind to playback without stopping the tape in between. The tape can also be badly stretched or broken by such an error. A logic-controlled system, on the other hand, keeps track of what it's doing at all times. If you hit the playback button while the machine is rewinding, the drive system automatically stops itself before switching into the playback mode.

Once logic control circuitry is included in a tape recorder, a number of nice (albeit nonessential) features can readily be added. Such features include remote control, memory rewind, and automatic tape reversal.

MONITOR SYSTEM

Your studio needs a monitor system so you can listen to what you are doing on the tape. For this you need some sort of amplifier and speakers. You can use your existing stereo system; however, this may only be practical if your studio is in the same room where you generally listen to your stereo. Lugging equipment back and forth can be a major nuisance.

You might feel that you can use cheap equipment because you are using the studio amplifier and speakers for monitoring, rather than sitting back and listening. If you're just using the monitor to find your place on the tape for editing, you might get away with this. If you want the best sounding tapes possible, though, you must use the monitor system to check the sound quality. If you overcorrect the tape to compensate for the limitations of the monitor system, the result is likely to sound quite bad when played back over decent equipment.

Ideally, the monitor system should be identical to the playback system. When this isn't possible, and it rarely is, the rule is to have a monitor system that is as good as possible. If anything, the monitor system should be better than the listening system.

How Much Power Do You Need?

There has always been a lot of controversy over how much power is enough. There seems to be no way to discuss this question without resorting to personal opinion. Feel free to disagree with me. I will try to present the issue, as I see it, as fairly as possible.

A lot of people are sold on the idea of as much power as possible. They like their music LOUD! They have a right to their tastes, but the ear does tend to distort at high volumes. You don't want any avoidable distortion when monitoring tapes. Whether it comes from the amplifier, the speakers, or the listener's ears, playback distortion prevents you from knowing what is really on the tape.

In an average-sized room, an average power of 2 to 5 watts should be loud enough, at least for monitoring purposes. Does this mean a 10-watt amplifier is sufficient? Not necessarily. Music is filled with brief peaks that are significantly higher than the average level. If the amplifier does not have enough dynamic headroom (extra power) to handle these peaks cleanly, clipping distortion results (Fig. 5-4). For an average power level of 5 watts, peaks of 50 to 70 watts are not at all uncommon.

There is a difference between electrical power and acoustic power, although both may be expressed in terms of wattage. Fifty watts of acoustic power would be completely unbearable, even for the most ardent rock fan. For comparison, consider the average

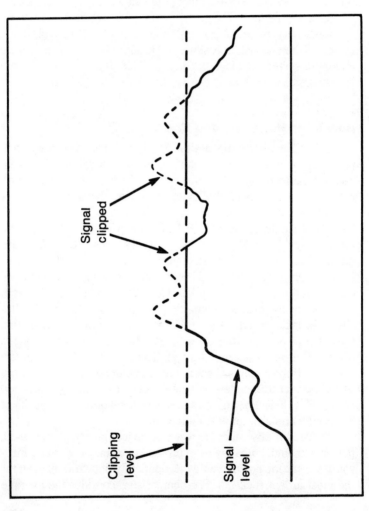

Fig. 5-4. Your monitor amplifier must have sufficient dynamic range to prevent clipping distortion.

acoustic power of the following sound sources:

conversational speech	0.024 mW	(0.000024 watt)
bass singer	30 mW	(0.03 watt)
saxophone	300 mW	(0.3 watt)
piano	400 mW	(0.4 watt)
pipe organ	13 watts	

Why is so much electrical power required? Because speakers are extremely inefficient. Here is a comparison of the typical power conversion efficiencies of some common types of speaker drivers:

acoustic-suspension speakers	0.2%
ported speakers	1%
horns	15%

The rest of the electrical power is wasted heating up the components in the speaker system. This is why applying too much power to a speaker can damage it.

Clearly then, the more power an amplifier is capable of, the better it can handle these peaks without distortion. (Curiously, many people who like their music loud sometimes prefer an amplifier with noticeable clipping to a cleaner sounding, higher-powered unit.)

In the home studio, a good rule would be to select a monitor amplifier rated for about 30 to 50 watts. More won't hurt, but may be wasteful. When shopping for an amplifier, an extra 10 watts shouldn't warrant much of a price increase, all other factors equal of course.

Amplifier power ratings can be misleading. All other factors being equal, a 100-watt amplifier might not sound appreciably louder than a 50-watt amplifier. To achieve a doubling of volume, you need to increase power by 10. To get twice the volume of a 50-watt amplifier, you need a 500-watt amplifier.

Amplifier power is an important specification, but it should not be considered the only or the most important specification. Some ads I've seen tell you the power rating and nothing else. Don't select an amplifier solely on that basis. Many other specifications are also important—frequency response, distortion, S/N level, input and output impedances, and others.

You should also consider the features the amplifier might have. Does it have built-in noise reduction? Headphone capability? How convenient and logical is the switch placement? Do you need multiple sets of speakers? Are there enough high-level (AUX) inputs? Is a provision for a microphone included? What does the amplifier have in the way of tone controls? The importance of these kind of things depends on your specific needs.

For a studio monitor, a tape monitor circuit (access to the preamp outputs and main amplifier inputs) is really close to essential. In a multideck studio, an amplifier with two tape monitor circuits would come in handy.

Monitor Speakers

Choosing an adequate amplifier is relatively easy. The specifications are as clear-cut and as meaningful as manufacturer's specifications ever are. Always take the specifications as a whole. One specification—such as power—is never a sufficient basis for making a decision.

Speakers, on the other hand, are practically impossible to provide meaningful specifications for. Take *all* speaker specifications with a large, economy-sized grain of salt.

There are three basic classes of speaker specifications:

- ☐ Impedance
- ☐ Power-handling capability
- ☐ Frequency response (range and degree of flatness)

Unfortunately, none of these are at all clear-cut. Speaker impedance might seem to be simple enough. If your amplifier is rated for 8 ohms (pretty much the standard today), just get an 8-ohm speaker. Right? Well, yes and no . . .

Impedance. Impedance is a complex term that varies with frequency. The actual impedance of a speaker depends on the frequency components of the instantaneous signal being applied to the speaker.

The rated speaker impedance is basically an average. It is often treated as a simple dc resistance for convenience in performing various calculations, but this can lead to serious problems if you take the numbers too literally.

Be careful playing games with impedance ratings. Connecting multiple speakers in parallel can lead to considerable damage to the amplifier and/or speakers. Unless you know precisely what you are doing, I would seriously recommend that you not do it. If you need multiple speakers, use an amplifier with multiple outputs or a specially designed speaker switch with an appropriate impedance-matching network.

Power Handling. Power specifications can be confusing for speakers. Ideally, two specifications should be given—a minimum and a maximum. Sometimes this is done, but not always.

The minimum operating power specification is sometimes given in terms of speaker efficiency, which is a measurement of how much power it takes to drive the speaker. Too little power applied to a speaker cannot overcome its mechanical inertia. The speaker cone will not move. Applying slightly more power will partially drive the speaker, but the sound will not be clear. Before the speaker starts reproducing a clean sound signal, X watts are required. Which condition does the minimum power specification define? It depends on the measurement.

Speaker efficiency indicates how loud the sound from the speaker will be for an electrical input signal of a given electrical strength. Unfortunately, there is no industry standard for measuring speaker efficiency, so ratings for speakers from different manufacturers (sometimes even from the same manufacturer) often can't be directly compared.

Most speakers are given a maximum power specification. This is usually the maximum continuous power the speaker can safely handle. In most cases, this measurement is made with a continuous single frequency sine wave signal. This does not correspond to what happens in music, but this method makes it easier to establish a comparison standard.

Generally, a speaker can tolerate brief peaks that are significantly higher than the continuous power rating. On the other hand, sometimes a speaker can be blown even if the continuous power rating is not exceeded. This can happen when the signal has a strong high-frequency harmonic content, when the speaker is overheated, or from other causes.

It is not a good idea to drive a speaker with an amplifier that has a higher power rating than the speaker—even at low-volume settings. The speaker could be damaged by a strong peak or if the volume control is accidentally turned up.

As a rule of thumb, the power rating for the amplifier should be 50 to 80 percent of the continuous power rating of the speaker. If the speaker is rated for 100 **watts,** a 50-watt to 80-watt amplifier would be your best choice.

If you regularly have the volume control setting at a position above its midpoint, perhaps you should get a higher-powered amplifier and speaker. If an amplifier has to work too hard, it may cause distortion and clipping, creating a strong spurious, high-frequency harmonic content that can damage the speaker even if the power rating is seemingly not exceeded.

If the amplifier's power rating is less than about 50 percent of the speaker's power rating, there might be a tendency to overdrive the amplifier. This might result in potentially damaging harmonic distortion.

The power-handling capability and efficiency of a speaker can vary considerably due to a great many variables, including temperature, the frequency content of the signal, and other factors. Thus, the power ratings for speakers are imprecise. They can be useful for comparison, but again, the numbers should not be interpreted too literally.

Frequency Response. The most important and confusing area of all is frequency response. In a very real sense, there is no such thing as a truly meaningful frequency specification for a speaker. That may be a little too harsh. Speaker frequency response ratings can be useful under some circumstances, but they almost never mean quite what they seem to mean.

Why is it so hard to define the frequency response of a speaker system? The reason is that the room the speaker is used in is actually part of the speaker system. Moreover, the placement of the speaker within the room can have a significant effect on the frequency response. There is absolutely no way to measure the frequency response of a speaker to get results that hold true under all typical operating conditions.

Many times I've encountered this situation: In one room, speaker A sounds unquestionably better than speaker B; but when both speakers are moved to a second room, speaker A sounds rather poor, while speaker B sounds great. I repeat, the room itself is part of the speaker system!

There are two basic types of frequency specifications for speakers—frequency range and frequency balance. Frequency range is simply the end points of band of frequencies the speaker can reproduce. For example, a typical speaker system might have a frequency range of 50 Hz to 20 kHz. This measurement is relatively constant, regardless of the environment. One potential problem area is defining the cut-off point. A speaker produces frequencies outside its defined frequency range, but at a lower level, as shown in the simplified frequency response graph in Fig. 5-5. When the level drops below a specific point, it is said that the fre-

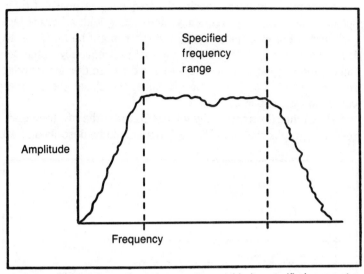

Fig. 5-5. A speaker reproduces frequencies outside its specified range at a reduced amplitude.

quency component has gone outside the range of the speaker. There may be some variation in what the cut-off point is.

Now comes the really tricky part—frequency balance. No speaker reproduces all frequencies equally. A real speaker frequency response graph is filled with irregular peaks and valleys, as illustrated in Fig. 5-6.

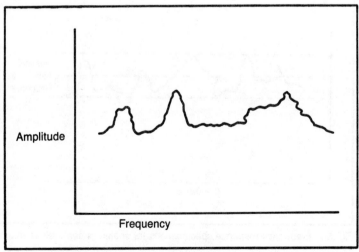

Fig. 5-6. No speaker gives really smooth frequency response.

Most manufacturers specify the frequency response of their speakers by giving the frequency range along with a "fudge factor" in dB. For example, a typical speaker might be rated for 40 Hz to 18.5 kHz ±4 dB. This means that no frequency in the defined band is greater than or less than 4 dB from the nominal flat response level. The lower the dB value, the closer the speaker comes to flat response.

How much does this really tell us? Consider the two frequency response graphs shown in Fig. 5-7. Both could be described as 40

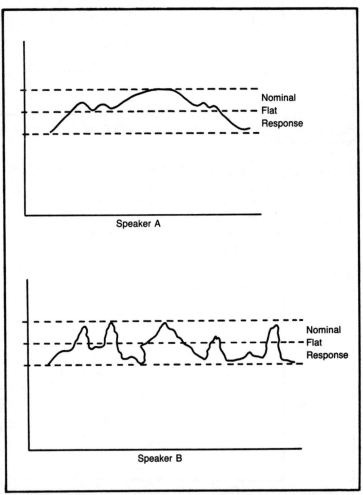

Fig. 5-7. Frequency response specifications for speakers don't tell as much as it might seem.

Hz to 18.5 kHz ±4 dB, but speaker A probably sounds a lot better than speaker B.

Can you judge a speaker by the frequency response graph given in some manufacturers specifications sheets? Not very reliably. The problem is that the frequency response graph is dependent on the specific manner in which the measurement is made. It might not correlate to the speaker's behavior under other circumstances.

Let's take a look at how frequency response is measured. Generally, a microphone with known frequency response characteristics is placed at a specific distance and at a specific angle with respect to the speaker. Then pure test tones running through the frequency range are fed to the speaker, and the amount of sound picked up by the microphone is measured and graphed. The known characteristics of the microphone are compensated for in making the graph.

This is a logical approach, and nobody has come up with a better method. Unfortunately, even this "best" method isn't very good. The problem is, as stated earlier, that the room itself serves as part of the speaker system. There is no way to separate the room's frequency response from the speaker's. Several speakers tested in the same room can be compared, but the comparison is really only valid for that room, or one very similar. In a different room, very different results might result.

For example, let's say some of the rooms dimensions cause a resonant emphasis around 7,000 Hz. (All normal rooms have several such resonant frequencies.) A speaker that is deficient around 7,000 Hz tends to sound better in this room than one that produces this frequency flatly or with slight peaking. The "better" speaker might not sound as good.

In addition, microphone placement can have a dramatic effect on the measurements. Especially important is whether the microphone is placed on-axis (pointed directly at the speaker) or off-axis (angled away from the speaker).

Figure 5-8 illustrates why all these problems crop up. Sound waves bounce off various surfaces in the room and are reflected back to the listener in extremely complex patterns. (This diagram is greatly simplified.) At any given position in the room some frequency components are in phase and emphasized, while others are out of phase and de-emphasized. Therefore, even if the speaker itself did put out a ruler-flat frequency response, which is theoretically impossible, listening to it or measuring it in the room would produce a very nonflat frequency response.

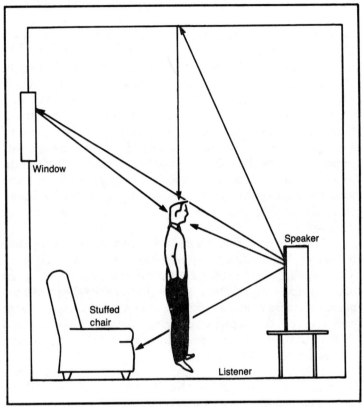

Fig. 5-8. The listening room is a major part of any speaker system.

Microphone placement can be standardized for measuring several speakers in a given room. The comparative results, however, would still only be valid for that particular room.

In many cases, speaker frequency response measurements are made in anechoic chambers. These are specially designed, acoustically-dead rooms with no echoes—or sound reflections—at all. Thus the effects of the room are neutralized. Only the speaker itself is measured.

On a theoretical level, this sounds like a good solution. Many experts feel the reality of the situation is quite different, however. The measurements taken in an anechoic chamber do not correspond with how good the speaker actually sounds in a real listening environment. I personally have encountered a number of speakers with great specifications made in an anechoic chamber that sounded only fair (or poor in a few cases) when I performed listening tests.

Consequently, frequency response specifications for speakers shouldn't be given too much weight. They aren't completely worthless—gross defects can be spotted—but never base the choice between two speakers solely on the frequency response specifications. Far more meaningful comparisons can be made by performing listening tests yourself—preferably in the room you intend to use the speakers.

In my opinion, the best specification sheet for a speaker would include four frequency response graphs in each of four standardized rooms. (The standards have not been defined at this time.):

- [ ] Small "hard" room (lots of reflective surfaces like glass or hard walls)
- [ ] Small "soft" room (lots of absorbing surfaces like drapes or thick carpeting)
- [ ] Large "hard" room
- [ ] Large "soft" room

The multiple frequency response graphs allow you to spot consistencies in the speaker's behavior.

Converting a frequency response graph made in one environment to meaningful data for a different environment is no easy task. To the best of my knowledge, nobody has ever done it. Even today's most powerful computers can't handle all of the variables involved. You can't really define the frequency response characteristics of a speaker. The environment has too significant an effect.

There is a lot you can do to compensate for the frequency response of a speaker, assuming it's of fairly good quality. For one thing an equalizer (see chapter 10) can help flatten out the response for your room. Speaker placement can also do a lot—especially in the bass region. A speaker placed near the floor or a corner emphasizes bass response. Absorbing materials, such as heavy drapes, overstuffed furniture, and thick carpeting can significantly cut down on the high-frequency response.

Most high-fidelity speaker systems include at least two drivers. It is highly unreasonable to expect a single speaker driver to accurately reproduce all frequencies in the audible range. Most quality speaker systems feature at least two drivers: a woofer and a tweeter.

A *woofer* is designed to reproduce low frequencies. To generate strong, low-frequency signals, the driver must move large volumes of air. This can be done with a large speaker cone that doesn't move far back and forth or with a smaller cone that moves further. (The design of the enclosure also has an important effect on bass response.)

High frequencies, however, require that the cone move back and forth very rapidly, hundreds or thousands of times a second. A woofer is too large and unwieldy for this. It moves too sluggishly. Portions of the cone surface tend to vibrate, resulting in very irregular response and directionality. A small driver called a *tweeter* is used to reproduce the high frequencies. Strong bass signals fed to a tweeter can damage it.

Many speaker systems also have one or more midrange drivers to extend the "division of labor" concept even further. Three or more drivers does not necessarily mean better response. Many two-driver speaker systems sound better than certain three-driver systems. With good design though, a three- or four-driver system can provide significant improvement in the sound quality. Just remember that this improvement is not an automatic result of adding more drivers to the system.

Headphones

Your monitor system should have provisions for headphones. You will often need to monitor a previously recorded track or some other sound source while recording a new track. If monitor speakers are used, the sounds will mix together on the new track. With headphones, you can listen and record at the same time.

It is not advisable to equalize a recording (see Chapter 10) while listening to it through headphones. The frequency response characteristics of headphones are very different from those for speakers. All too often a mix that sounds great through headphones, sounds awful when you play it back through regular speakers.

There are two types of headphones. Some are designed to be audio transparent; that is, external sounds can be heard while listening through headphones. Other headphones are designed to completely cover the ears and block out external sounds as much as possible. For monitoring while recording, the sound-blocking type are preferable. You can listen to what is going on the tape without being confused by any extraneous sounds.

IMPEDANCE MATCHING

When interconnecting various pieces of electronic equipment, the input and output impedances should be properly matched. All electronic devices, including passive devices such as cables, microphones, passive mixers, loudspeakers, etc., have characteristic input and output impedances, which are often written as Z. The signal goes from the output of one device to the input of the next.

Electronics theory states that maximum power transfer is achieved when the output and input impedances connected together are equal. In audio work, however, other factors become important. Maximum power transfer is not the most important consideration.

Theoretically, the output impedances should be as low as possible. A low output impedance offers a low resistance to the passage of the signal. This means the output can drive multiple connections without a loss of performance or a significant voltage drop in any portion of the signal path. There are practical limits to how low the output impedance can be. These are defined by the power supply's capabilities and the characteristics of the components in the circuitry.

Input impedances, on the other hand, should be as high as possible. A high input impedance implies that the circuit does its job with a minimum of electrical energy as a beginning. This helps limit loading of the preceding stage.

Matching impedances can thus be a fairly tricky proposition. Many technicians use a 7:1 ratio as a guideline; that is, the input impedance should be at least seven times the output impedance.

To make matching more convenient, many manufacturer specification sheets list the output's advised load impedance. If the suggested load impedance of an output is 10,000 ohms, the signal should be fed to a 10,000-ohm input. It is much easier to match impedances using the load impedance method. It is easy to get confused, however, because there are two ways to list the advised load impedance for an output:

- ☐ Minimum load impedance = 10K
- ☐ Maximum load impedance = 10K

The two specifications might appear to be exact opposites, but actually they are saying precisely the same thing. It's just the awkward wording used in the specification sheets that lead to confusion.

Minimum load impedance is the term used to define the minimum advised impedance of the load. In the example, the load impedance should not be less than 10K (10,000 ohms). It might be greater than this value, but not less.

Maximum load impedance refers to the impedance of the maximum advised load. Remember, a high impedance input puts less of a load on the source output than a low impedance input. The greater the load, the less the impedance. In the example, the impedance of the maximum acceptable load is 10K (10,000 ohms). Because increasing the load decreases the impedance, the load impedance should not be less than 10K, although it may be greater. Despite appearances, the two seemingly contradictory specifications are identical in meaning.

Both forms of this specification are in widespread use. It would be nice if everybody would agree to one standardized wording, but it isn't likely to happen in the near future. Just remember that minimum load impedance is the same as maximum load impedance. This is one of those times where you have to ignore your common sense and accept the ridiculous because that's the way it is.

If you are connecting one input to one output, matching the impedances shouldn't be too much trouble. But what if you want to drive two (or more) inputs with a single output using a "Y" connector, as illustrated in Fig. 5-9? The parallel impedances must be calculated to find the actual load impedance seen by the output.

If all of the input impedances are equal, simply divide the impedance of one input by the number of inputs. For example, two 10K impedances present a combined load impedance of:

$$10,000/2 = 5,000 \text{ ohms} = 5K$$

Or, let's say you have three 21K (21,000 ohms) input impedances paralleled together. In this case the combined load im-

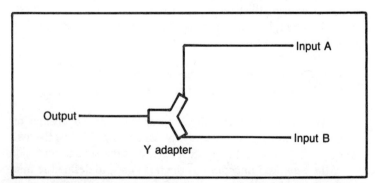

Fig. 5-9. If two inputs are to be connected to a single output via a Y-adaptor, the parallel impedance of the combined inputs must be considered.

pedance would be equal to:

$$21,000/3 = 7,000 \text{ ohms} = 7K$$

Suppose you want to parallel two unequal impedances, say 12K and 35K. Take half of the smaller value (in the example this would be 12K/2 = 6K). If this is greater than the advised load impedance of the output, you are safe.

For a more exact value for the combined load impedance, use this formula:

$$Zx = (Z1 \times Z2)(Z1 + Z2)$$

Using the values in the example, you find that the combined load impedance works out to:

$$Zx = (12,000 \times 35,000)/(12,000 + 35,000) =$$
$$420,000,000/47,000 =$$
$$8936 \text{ ohms} = 9K$$

If there are more than two impedances to be combined, this formula can be used:

$$Zx = 1/((1/Z1) + (1/Z2) + \ldots + (1/Zn))$$

As an example, assume you are paralleling the following four impedances:

> 10K
> 35K
> 22K
> 50K

The combined load impedance in this case works out to:

$$Zx = 1/((1/10,000) + (1/35,000) + (1/22,000) + (1/50,000) =$$
$$1/(0.0001 + 0.000029 + 0.000045 + 0.00002) =$$
$$1/0.000194 =$$
$$5150 \text{ ohms} = 5K$$

Remember the combined load impedance should never go below the specified minimum load impedance (or maximum load impedance) for the output being used as the signal source.

STUDIO ACOUSTICS

The room you record in has an important role in the sound and quality of your recordings. Just as the room's characteristics can act as part of a speaker system, it can create noticeable effects on the sounds picked up by a microphone.

Ordinarily you won't be aware of room effects, unless they are severe. The subconscious tends to compensate for the audible effects of the environment. So why can't most room effects be ignored in recording? Because the subconscious won't cooperate!

When listening to a tape, there are two possibilities—either you listen to the tape in the same environment in which the tape was made or you listen to it in a different environment. Now, consider what happens in each of these cases:

- ☐ Same Environment—The room effects occurred during recording, then again during playback. In other words, the room effects are doubled. The exaggerated room effects could easily be too strong for the ear to ignore. Not good.
- ☐ Different Environment—The subconscious adjusts the ear for the current environment. It can't compensate for room effects on the tape too. Consequently the recorded room effects are audible. Not good.

Clearly then, room effects cannot be safely ignored during recording. Occasionally you might get away with it, but don't count on it.

The remainder of this chapter discusses some of the most important acoustic factors involved in recording environments. Some suggestions on how to deal with these problems are also presented. This is not an exhaustive treatment of the subject, but it should be sufficient to get you started.

Soundproofing

Noise, in the sense of unwanted sounds, is the most obvious effect of the recording environment. Soundproofing is a vital consideration in any recording studio.

Soundproofing is important in two directions. You certainly don't want sounds from outside the studio to get into your recording. On the other hand, especially in home studios, you don't want sounds from within the studio to get out, or you'll be faced with angry neighbors and/or family members.

The most direct form of soundproofing is to stop the noise at the source. If the studio is located in a quiet area, and if you can get the rest of your family and neighbors to cooperate, many noise problems can be eliminated before they occur. Ask that the TV be turned down while you're recording. Unplug the phone.

If there is an apartment above your studio, the best way to eliminate the sounds of footsteps from above is to get your upstairs neighbor to lay down carpet. It may even be worth the expense to offer to pay for the carpeting yourself. It would probably be cheaper (and much simpler) than to try to block out the sound at your ceiling.

Regrettably, most external sounds simply can't be eliminated at the source. Life persists in going on around us. Some neighbors or family members may not see any reason to cooperate, or they might forget while you're in the middle of an important recording session. Then there are the outside sounds you have no control over, such as traffic noise, planes flying overhead, etc. Even if you set up a studio in an isolated cottage miles from civilization you will still be faced with sounds from outside—birds chirping, thunder, gusts of wind, animal cries, etc.

The conclusion from all this is inescapable. All recording studios *must* have some kind of soundproofing.

A wide variety of approaches to soundproofing are available. They range from the extremely elaborate and expensive to relatively simple and inexpensive tricks.

I will start with the ideal approach and work my way down. I doubt if many readers can afford the techniques described first, which are generally used in professional studios, but knowing how the pros tackle the problem can be informative. The less expensive techniques described later are clearer if you first understand the direct approach.

One of the best ways to soundproof a room is to isolate it from adjoining rooms, as illustrated in Fig. 5-10. The air between the outside and inside walls acts like a very efficient "sound trap." Entrances should be via sound locks (similar in principle to an air lock on a spaceship), as shown in Fig. 5-11. Notice also that the floor is "floating," or supported on springs or rubber pads that act as shock absorbers. This ideal studio is also completely airtight.

This plan is not completely soundproof. If, for example, a demolition crew is blasting in a building down the street, you would probably hear some of the noise in the studio. This type of studio comes

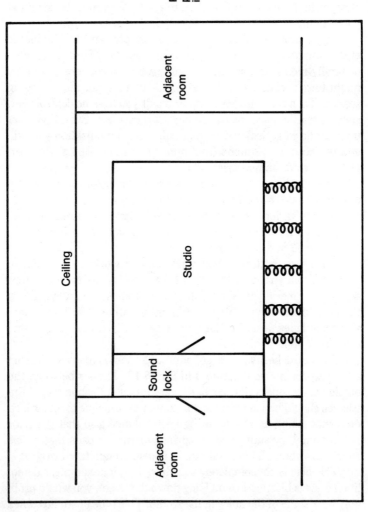

Fig. 5-10. The best way to soundproof a room is to completely isolate it.

very close to perfection though in terms of soundproofing. It would be more than adequate in at least 99 percent of all recording sessions.

Unfortunately, this kind of structure is only practical if it is constructed from scratch. Moreover, it is very, *very* expensive. Even the majority of professional recording studios have to make compromises to this scheme. The hobbyist or semiprofessional has no hope of achieving such a high degree of soundproofing.

Fortunately, good—if not excellent—soundproofing can be achieved for much less. The idea is to use the principles exhibited by the ideal design, but implement them in less expensive ways.

There are three basic types of soundproofing:

- ☐ Blocking
- ☐ Isolation
- ☐ Absorption

All three are tightly related, and there is a lot of overlap between the categories. The only point to separating them in this discussion is to help clarify how some of the methods for soundproofing work.

The ideal studio is airtight. No outside air gets into the recording environment. (Presumably some kind of air conditioning/ventilation system is included to prevent suffocation.) Since most sound waves travel most easily on air, blocking the air blocks out much of the sound.

A 1-square-inch opening can allow as much sound to get in as an average solid wall. Blocking off as many air openings as possible goes a long way towards soundproofing a studio. Plug up any cracks or holes, even in inside walls. Use weatherstripping to block the cracks around doors and windows. No doorless entrances should exist in the studio. Put up a tight-fitting solid wood door, or at least a partition of gypsum board, or plywood. Block off every opening you can locate.

Just the weatherstripping around the cracks of a door can improve the outside noise rejection about 3 to 10 dB. That is definitely worthwhile and shouldn't cost more than a reel or two of tape.

Solid doors should be used whenever possible. Hollow doors are cheaper and lighter, but they don't offer much resistance to sound waves. These hollow doors are typically made up of two, 1/8-inch panels connected to a frame, with an air gap between them. A hollow door is certainly better than nothing, but a comparable

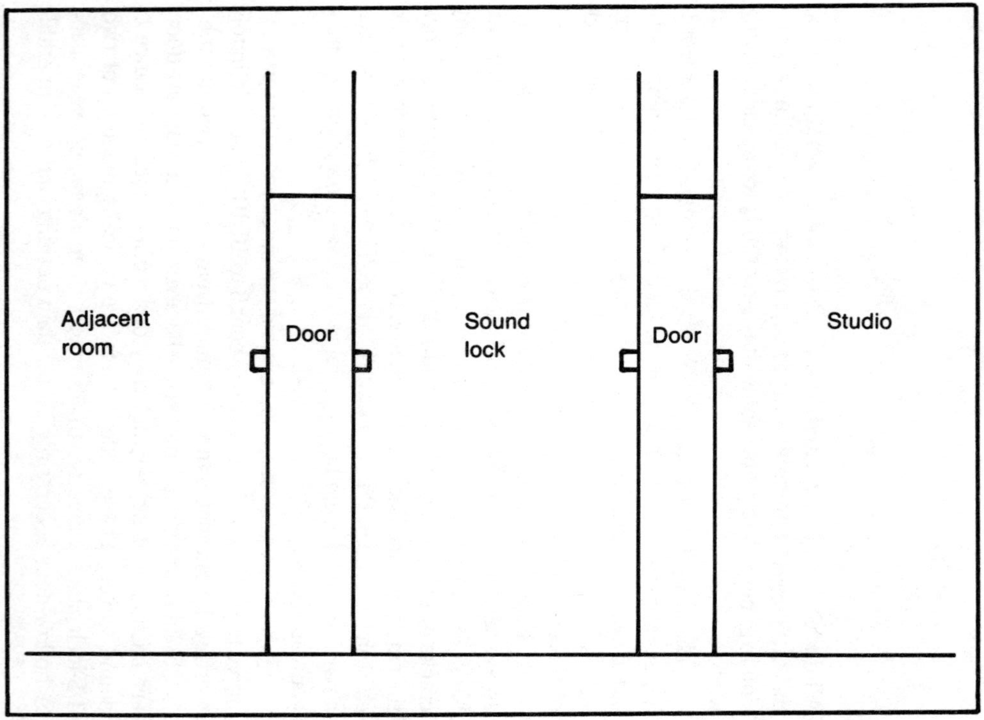

Fig. 5-11. Entrances to an ideal studio should be through "sound locks."

solid wood door offers an extra 10 dB or so of noise rejection. Used wooden doors can usually be purchased fairly cheaply.

Storm windows are also highly desirable in a recording studio. The dual panes of glass block outside sounds in much the same way they help block heat transfer.

Whenever possible pad or carpet the *outside* of your studio. The pad or carpet absorbs many sound waves before they get to the wall. Usually this is only practical if your ceiling is a floor for someone above you. Carpeting your upstairs neighbor's floor can solve a lot of noise problems in your studio.

It might seem logical that if you can't pad or carpet the outside, you could carpet or pad the inside surfaces of your studio to absorb sound waves after they pass through the wall (floor, ceiling). It might sound logical, but it doesn't work. Internal padding might cut incoming noise down 2 to 5 dB, but not an appreciable amount. And it can do a lot of damage to the sound quality within the studio. Internal absorption makes the studio sound quite "dead." This is discussed shortly.

If possible, hang a suspended ceiling in your studio. The air space between the studio's ceiling and the floor above should be lined with fiberglass or a similar absorbent material.

If you do have a suspended ceiling in your studio, heavyweight ceiling tiles do a better job than lightweight tiles. Such ceilings are often constructed with so-called acoustical tile. This type of tile is absorbent, but not insulative. It helps absorb sounds within the studio (deadening the sound quality), but does not insulate against outside sounds. Such acoustical tiles are not desirable in a recording studio. Heavier tiles made of fiberboard, gypsum, or plywood are a much better choice.

Other noise problems might be generated within the studio itself. Loose objects tend to rattle when a loud sound is produced in the studio, especially when the sound is near the resonant frequency of the loose object. This can result in objectionable humming and buzzing.

The solution to this kind of problem is obvious enough. Remove such loose objects from the studio or secure them solidly. The kinds of objects to watch out for include ash trays, bric-a-brac, dishes, glasses, soda bottles, lamps, paintings and other wall hangings, statues, and small tables.

Appliances such as heaters and air conditioners can be very noisy. Any nonessential appliances should be turned off during recording sessions, though this is not always possible. In the winter, for example, it may be highly undesirable to turn off the heater for long periods of time.

The only thing to do in such cases is to isolate the appliance(s) from the acoustic environment as much as possible. The equipment should be mounted on a rubber pad or springs. This helps prevent vibrations from the appliance from spreading through the studio.

Enclosing the appliance in some kind of soundproof housing also helps cut down undesirable sound transmissions. Ideally, from an acoustic point of view, the enclosure should be airtight, though this simply wouldn't be feasible in many situations. A heater enclosed in an airtight container won't operate properly and might even be a fire hazard. Use common sense in isolating appliances. It is better to accept a little noise than to set up a potentially dangerous situation.

Remember, you will never get rid of all the noise in any environment. If nothing else, random movement of air molecules is a form of noise. Your recording equipment also introduces some noise into your recordings.

While you can't achieve 100 percent soundproofing, you should be able to reduce the environment's noise level to an acceptable level by using the methods described in this section.

Room Acoustics

In the section on speakers, you learned that the characteristics of the room itself can act as a literal part of the speaker system. Similarly, the acoustic characteristics of your studio have a considerable effect on your recordings.

Even the most directional microphone available picks up sound from many different parts in the room. In addition to the actual sound source being recorded, the microphone also picks up multiple reflected sound waves from various surfaces in the room.

Hard surfaces, such as glass or solid walls, reflect most of the sound energy. Softer surfaces, such as heavy drapes or carpeting, absorb most of the sound energy striking them.

A recording made in a highly-absorbent environment sounds very flat or "dead." A too reflective recording environment results in a muddy, indistinct sound. Between these two extremes lie a variety of sound qualities that might or might not be desirable for different types of recordings. In an ideal studio you have full control over how "live" or how "dead" the sound is. Full control isn't practical, but you can achieve many different effects without too much trouble.

Before you start treating your studio acoustically, make a number of test recordings. The room might be fine as is. Make test recordings after each stage of your acoustic treatment. When the recordings sound good, stop. There is no sense in going to unnecessary effort and expense. Overdoing acoustic treatment can actually degrade the sound of your studio. Do things one step at a time.

Breaking Up Standing Waves. Parallel reflective surfaces (especially large walls) should be avoided in the recording studio. The repeated echoing back and forth of sound waves between the surfaces could lead to *standing waves*. Certain frequencies will be partially or even almost completely cancelled out in some parts of the room and overemphasized in others. It is hard to achieve good microphone placement in a room with strong standing waves.

The solution to this problem is to just break up one or both of the parallel surfaces with patches of absorbent material. Patches of plush carpeting, cork, or acoustic tile irregularly placed on a wall can do wonders in decreasing standing wave effects. Acoustically the wall then appears to be a number of irregular surfaces instead of one large reflective surface. Even painted egg cartons placed on the wall can create an irregular surface to break up standing waves.

If patches of plush carpeting are used for this purpose, be very careful not to overdo it. Thick carpeting tends to absorb high frequencies. Without any reflection echoes, the high frequencies sound "dead" while the low frequencies are still reflected for a "live" effect. This combination sounds unnatural and makes a poor recording.

Basstraps. Absorbent patching does a lot to cut down standing waves and room *coloration*. In many cases, however, additional measures might be required, especially for very low-frequency standing waves. Constructing a few simple *basstraps* would probably help.

A membrane absorber-type basstrap is also sometimes known as a *resonator*. A box-like frame contains an air cavity. A rigid membrane of plywood or linoleum for instance, is stretched over the frame.

The membrane absorber has a fundamental frequency that is determined primarily by the flexibility and weight of the membrane material. When a sound wave near this fundamental frequency

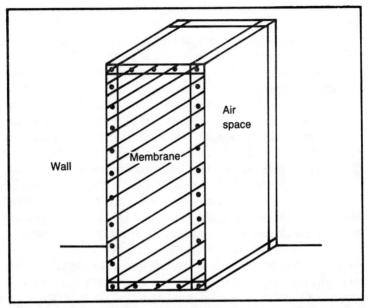

Fig. 5-12. This simple membrane absorber basstrap can be used to minimize room coloration.

strikes the membrane, it starts to vibrate and sets the air in the cavity behind the membrane into motion. Both the "damping" motion of the stiff membrane and the resistance of the air cavity behind it combine to use up (by absorption and dissipation) the energy of the sound wave. This device is also resonant at the lower harmonics of its fundamental frequency.

The effective range of frequencies can be extended by lining the internal air cavity with a porous, absorbent material, such as fiberglass. This gives a smoother response from the resonator. A typical frequency response graph for a resonator without a lined air cavity is illustrated in Fig. 5-13. A frequency response graph for a similar resonator with a lined air cavity is shown in Fig. 5-14. A few well chosen basstraps can go a long way towards equalizing the frequency response of your studio.

Floors. Experts are in disagreement over whether the floor of the studio should be carpeted. On the one hand, carpeting tends to absorb vibrations, especially from noisy instruments such as drums. On the other hand, carpeting also tends to absorb high frequencies. If carpeting is used, it should be a fairly coarse weave. Do not use a plush carpet on the floor of your studio. It is too absorbent.

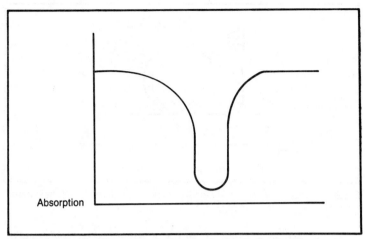

Fig. 5-13. An unlined resonator has a very narrow active band.

Floating Platforms. Noisy instruments like drums can also be isolated from the studio floor (limiting unwanted vibrations) by setting them up on "floating" platforms (Fig. 5-15). These platforms don't have to be particularly elaborate. A sheet of 1-inch plywood over a 3-inch layer of soft foam does nicely in most instances. For even better isolation, try a 6-inch to 8-inch deep box filled with sand with a thick sheet of plywood over the top. The box itself should be made of fairly heavy plywood or a similar material.

Microphones. In setting up for recording, the microphone(s)

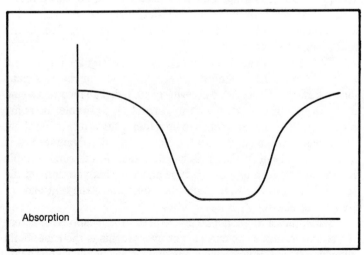

Fig. 5-14. Lining the inside of a resonator can give a broader response band.

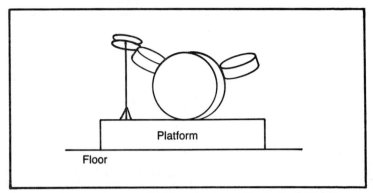

Fig. 5-15. Noisy instruments should be set up on "floating" platforms to prevent vibrations in the studio.

should be placed as far from all major reflective surfaces (walls, floor, ceiling) as possible. The exact center of the room would be the best choice on this basis, although it isn't always a practical one.

Isolation Systems. Reflections from the nearest major reflective surface clearly have the greatest effect, so that surface should be given the most attention when acoustically treating the studio. When you can't get far enough from one of the major reflective surfaces, some kind of isolation system is needed.

The least expensive approach is the *tent*, which is illustrated in Fig. 5-16. Two lengths of clothesline are hung across the studio, 6 feet apart and 5 or 6 feet above the floor. Now a carpet about 18 feet by 5 feet is draped over the clotheslines, as shown in the diagram. If you want near "dead" (flat) response, the fuzzy side of the carpet should be on the inside of the tent. The floor of the tent should also be carpeted.

This tent is a simple and inexpensive method for minimizing acoustic room effects. Unfortunately, the space for the musicians (or whatever) to be recorded is severely limited. The recorded sound tends to be quite dead, which might not be desirable for many recordings. The tent also tends to be rather awkward to move about to various locations in the studio. Despite it's disadvantages, it is so simple and cheap that it is worth trying. Make some sample recordings within a tent of this type and see what you think of the sound. If it doesn't sound good, then take down the tent and try something else.

Sometimes it might not be practical to stretch clothesline across the studio to drape the carpet tent over. If this is the case, build a wooden frame. If you have a few extra microphone stands avail-

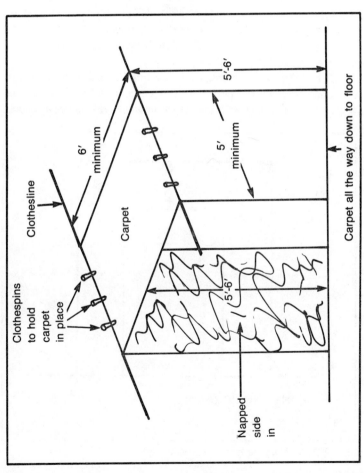

Fig. 5-16. A simple "tent" can be used as an inexpensive isolation chamber.

able, they can serve as a simple frame for a temporary tent.

If you decide on a permanent tent, you should consider doubling it, as shown in Fig. 5-17. The air space between the two carpets improves the effect. For maximum benefits from the double tent, the air space between the carpets should be 3 to 6 inches.

A three-sided folding screen, like the one shown in Fig. 5-18, can come in handy as a portable isolator. If two-way hinges are used, one side can be covered with carpeting for a relatively "dead" effect, and the other side can be left bare for a relatively "live" effect. The angle between the sides can vary within a limited range for different effects. Cover the spaces between sides with fabric that allows the hinges to function, but blocks off the air gaps that could neutralize the effect of the isolation screen.

Do not make this folding screen too heavy. The main point of this type of construction is portability. It doesn't make much sense to make it too heavy to move easily. More importantly, if the screen is too massive, it will resonate at low frequencies, creating standing wave problems of its own.

Portability can be increased by putting the unit on casters. The air space between the bottom of the screen and the floor can significantly reduce the isolation effect, however. If casters are used,

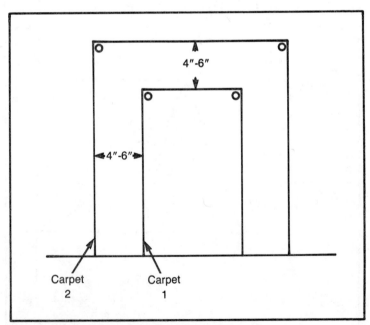

Fig. 5-17. Doubling the tent improves its performance.

add a double skirt of thick carpeting material along the bottom.

Portable isolation screens also come in handy for simultaneously recording several instruments on different tracks for later mix-down (see Chapter 9). You don't want the sound from one instrument picked up by another instrument's microphone. Isolating each instrument with folding screens minimizes this type of problem. The screened-off areas can be arranged so that the musicians can maintain eye contact. Most musicians give a better performance when the group feeling is preserved.

Absolute isolation between instruments is not possible. Fortunately it is rarely needed, and in many cases, it might even be undesirable. Multitrack recording is discussed in Chapter 9.

Another type of portable unit that comes in very handy in the recording studio is the *gobo*. A gobo is a sound isolator panel. Several gobos can be arranged to create a temporary acoustic *room* around individual instruments for multitrack recording. The sound waves from each instrument is more or less isolated by the gobos.

A fair gobo can be made by constructing an enclosed box of plywood, masonite, or particleboard and filling it with fiberglass or blankets. The heavier a gobo is, though, the better it works. The best gobos are solid, made up of several sheets of one of the materials previously mentioned in sandwich fashion.

One side of the gobo may be covered with an absorbent material, such as carpeting, to act as a sound absorber. For a more "live" sound, use the bare side of the gobo.

Reverberation. A *live* environment has numerous small echoes that enrich the sound and give it body. This is why it sounds so much better to sing in the shower. The sound is reinforced by bouncing off the hard tile. These echoes are called *reverberation*. Too much reverberation, however, can be a bad thing. The sound becomes muddy and indistinct.

A *dead* environment, on the other hand, has little or no reverberation. It sounds dull and lifeless—which is why it's called "dead."

A number of devices for artificially adding reverberation to recordings are available. While some are better than others, I've yet to find one that sounds as good as a well-designed performance hall. Still, they are better than nothing if your studio is too dead. Moreover, most reverberation units allow you to continuously adjust the amount of reverberation. Different types of music sound better with different degrees of "liveness."

Here's a simple trick that can give you some dynamic control over the natural reverberation in your studio. Hang a thick set of

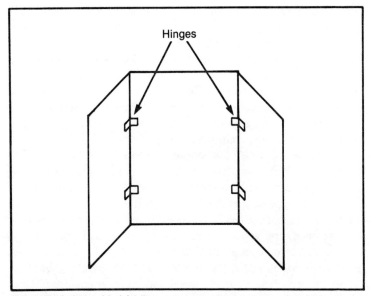

Fig. 5-18. A three-sided folding screen can be used as a portable isolator.

drapes over a solid, acoustically reflective wall. Closing the curtains gives maximum "deadness." Opening the curtains fully gives maximum reverberation ("liveness"). Intermediate degrees of reverberation can be achieved by partially opening the drapes.

The recording environment plays an extremely important role in the quality of your recordings. It deserves at least as much thought and effort as selecting your equipment (which deserves a lot). A poor studio seldom, if ever, gives satisfactory results. As this section has pointed out, there is a lot you can do to improve the acoustic performance of whatever room you are using as a recording studio.

Chapter 6

Live Recording

Sooner or later most recordists want to do live recording at some location other than the studio. This introduces new complications. Fortunately, they are not insurmountable.

One of the major problems with out-of-the-studio recording is that you generally have much less control over the environment. Noise and room acoustics are usually considerably different from in your studio. Portable gobos and isolation panels can help some, but your options are limited by the location you are recording in.

For live recording, an equalizer is essential to help compensate for the acoustic characteristics of the environment. Equalizers are covered in Chapter 10.

EQUIPMENT FOR LIVE RECORDING

Great care must be taken in selecting equipment for use in location recording. To be portable the equipment should be relatively lightweight and sturdy. Of the two, sturdiness is unquestionably the most important. You can always get a friend to help you carry heavy equipment; but if a machine is dropped, as it inevitably will be sooner or later, it is vital that the damage be minimal, or you're out of luck. Even without gross abuse such as dropping, just the wear and tear of repeatedly moving the equipment about can soon lead to problems if it is not designed to withstand such use.

For live recordings that do not require much editing or post-production mixing, you might want to consider a cassette deck. For field recording their advantages outweigh their disadvantages. They tend to be more portable and relatively lightweight than comparable reel-to-reel machines. The convenience of not having to thread tapes could also be an advantage. If your home studio and mixing equipment (see Chapter 9) are already set up for reel-to-reel equipment, however, changing formats for live recording might be more of a nuisance than a convenience.

Whenever possible, bring along plenty of spares (of everything) when going into the field for a recording session. You never know what unanticipated disaster may strike. It's smart to be prepared for just about anything. You probably can't afford to carry around spare tape recorders, mixers, and equalizers. These things are too expensive and bulky to make spares practical. You definitely should have a field recording kit, though, with plenty of spare accessories and parts for emergency repairs.

If you know how to do electrical repairs, bring along a few spare components that might be likely to go bad along with the appropriate tools—screwdrivers, pliers, soldering iron and solder, hook-up wire, etc. A small VOM (volt ohm-milliammeter) would also be a good idea. If you don't know your way around a circuit, don't bother bringing along these items. If you don't know what you're doing, you should never open the case on any piece of electrical equipment—you may do more damage. Refer all repairs to a trained technician.

If you can possibly afford it, two or three spare microphones would certainly make your life easier. Microphones are relatively delicate and might go bad after being moved from location to location. Spare microphone stands are also a very good idea.

Now, here are some things which should definitely be in your field recording kit. Bring along several extra reels of tape and some spare take-up reels. A splicing kit should also be included in your live recording kit. You need at least one set of headphones for monitoring as you record. An extra set of headphones is a good idea too. They don't have to be a super-deluxe type—just so they're good enough for you to hear what you're doing as you adjust the controls on your equipment.

Always bring along plenty of extra interconnecting cables. It's been my experience that the most frequent source of problems in live recording sessions comes from bad cables. Continuously rolling and unrolling cables and laying them out in various configurations puts them under considerable strain, breaking the tiny internal wires.

Cables should *always* be neatly, but loosely, rolled up for storage. A small twist tie or rubber band can hold them tidily. Never just toss cables haphazardly into a box. Besides the time you'll waste untangling them the next time you want to use them, this kind of abuse greatly increases the chance of internal breaks.

Your live recording field kit should also include every possible combination of adapters for going from one type of connector to another. You never know what you might have to plug into what.

Bring along several electrical extension cords of various lengths. You probably won't know until you get to the recording site where the electrical outlets are. You don't want to be forced to set up your equipment in an awkward location. Extension cords aren't expensive; there is no excuse for not having a few on hand just in case.

Don't forget several spare fuses for each piece of equipment. If a fuse blows and you don't have a suitable replacement, the entire recording session might be wiped out. Never replace a fuse with one that has a higher rating. The fuse rating is not selected arbitrarily. If you substitute a higher rated fuse, there is a good chance some of the circuitry will blow to protect the fuse. If the correctly rated fuse repeatedly blows, something is wrong. Unplug the equipment in question and get it to a competent repair shop as soon as possible. Don't try to cheat with a higher rated fuse to use the equipment, even for "just a few minutes." The odds are very, very strong that you'll just end up with a larger repair bill. *It is not worth the risk!* I repeat—*never substitute a higher rated fuse!*

A little bit of advance planning before you go to the recording location will save you a lot of grief, frustration, and wasted time. And if something goes wrong, and you didn't bring along the necessary spares, you've got nobody to blame but yourself. A field recording kit like I've described here is unquestionably worth the modest expense and extra trouble of setting it up and carrying it around. Problems crop up in field recording a lot more often than you might think.

SETTING UP

Always arrive well ahead of time for any location recording session. You would be surprised to realize how much time setting up your equipment can take. Estimate how long you think it will take you to set up, assuming that several problems will crop up. Then add about 50% to this. To be safer, tack on an extra half hour as a "fudge factor." Arrive at the location that much time ahead of the scheduled event to be recorded. If you're lucky, you'll have enough time to set up without rushing or compromising. The cardinal rule in field recording is: always expect the unexpected.

It is very important that you are familiar with your equipment and its capabilities before you go out on a field recording session. You won't have any time to experiment to find out if the equipment can do "such and such." Make several test recordings at home before the live recording session, experimenting with the equipment then. When you're setting up for live recording, you will be too busy trying to compensate for the location to experiment with the equipment.

Pay particular attention to microphone placement. Make as many test recordings as you can. Since few locations (including most concert halls) are really well-designed acoustically, your microphone placement will probably be a compromise, but you want to get the best compromise you can get. There are so many potential variables in microphone placement, so I really can't give any rules, except—experiment, experiment, experiment. Eventually you should be able to develop a "feel" for your equipment and develop some rules of your own, especially if you frequently record in the same location.

Set up your equipment in a convenient, usable manner on a table with a comfortable straight-back chair. Your setup should be out of the way of the performers and audience, if any, but where you've got a good view of what's going on. If lighting is not too good, especially in theatres where the lights will be dimmed, lighted panel meters or LED readouts are a big help.

If your setup is where any audience member can get at it, never leave the equipment unattended—not even for a minute. Theft is a sad fact of life, and a lot of thieves are very fast and very good at it. Even without the consideration of theft, all that impressive looking equipment can be an irresistible temptation for some people. They just can't keep their hands to themselves. They might innocently change a crucial control setting, unplug a cable, or even damage some of the equipment. If you have to leave the equipment table for any reason, post a reliable guard.

In setting up for a live recording session, a knowledgeable assistant will make your life easier. Don't just draft a friend who has never touched a recorder before. You'll waste more time showing them what to do or correcting their mistakes than you would if you just went ahead and did the job yourself.

No special knowledge is required for someone to just help you carry equipment or say "testing, testing" into a microphone. If you need help, you should be able to get it. But bring your own help, especially for live performances. Anyone who is there for the show will be busy enough—don't count on somebody having a spare moment or two to lend you a hand. Arrange for your assistant(s) ahead of time. Only ask someone you know will show up on time and actually work. A goof-off will just get in the way, and you'd be much better off without him.

Chapter 7

Care of Tape and Recorders

You can spend a small fortune on recording equipment, but if you don't use proper maintenance procedures, you're practically throwing your money away. You certainly won't get the performance you paid for very long. With less expensive equipment, good maintenance is even more important.

Far too many people don't bother with maintenance. Even those that do often do a quick, half-hearted job. It doesn't take that much time and effort to do it right, and you can save yourself a lot of money (in repairs and replacing equipment and supplies) and frustration (due to poor recordings or damage to tapes). There is no excuse for not setting up a regular maintenance program. In the long run, it will more than pay off in improved results from your equipment.

TAPE STORAGE

Many tapes have been destroyed due to improper storage. If your recordings have any value to you at all, develop good storage habits.

Always store the tape in an individual box. Accumulated dust and grime can cause lots of problems. The box also gives you ample space for noting the contents of the tape. Some tape manufacturers print a lot of information and fancy graphics all over the box, leaving little or no room for documentation. Just put the indexing information on a 3-×-5 card and tape it to the outside of the box.

This can also be done if you are reusing a tape from a previously marked-up box.

(This tip might sound rather trivial and obvious, but on several occasions when I have mentioned it to other recordists, many hadn't thought of it and considered it a pretty neat idea. So I'm passing it along to you.)

Tapes should never be stored in extreme temperatures. Excess heat is more of a potential problem than excess cold. Do not store your tapes near a heating vent or in direct sunlight. Also don't place them too close to any electronic equipment that might run hot.

It's even more crucial not to store your tapes anywhere near a magnetic field. If you do, your recordings may be erased. Do not place tapes on top of speakers (which have large, permanent magnets) or on top of a television set (the electron beam is deflected magnetically). If there is any question whether the proposed storage location is safe, pick a new location. Better safe than sorry. tapes should be stored in an upright position if possible. This is not terribly crucial, but it is advisable.

If tapes are wound too tightly, *print-through* problems might occur. A strong signal recorded on one layer of the tape might seep through and imprint itself on adjacent layers of tape. This can cause audible echoes and/or an overall muddiness in the sound during playback. Once print-through has occurred, the only way to correct it is to erase the tape entirely and redo the recording from scratch. Clearly this is often impossible, so it is important to avoid print-through.

Never store tapes just after they have been rewound or fast-forwarded onto the storage reel. These high-speed modes tend to wind the tape somewhat unevenly and too tightly, leading to possible print-through. Always load the tape onto the storage reel at playback speed.

Professionals often store tapes "tail out." This means the tape is wound on the take-up reel. It is rewound immediately before playback and stored without rewinding. For tapes recorded in both directions, there is not "tail out" or "head out." Never put too much strain on a tape. It can easily be stretched, ruining the recording.

TAPE HANDLING

Care should be exercised whenever handling magnetic tape. Oils from your fingers, not to mention dirt or grime, can adhere to the magnetic coating, contaminating the tape and possible damaging the recorder's heads the next time the tape is used. Many recordists always work with tape while wearing thin, lint-free gloves. While not essential, this is not a bad idea.

It is not a well-known fact, but the human body exerts a magnetic field. Usually this magnetism is at a negligible level. But some people, at some times can put out a fairly strong magnetic field. (This is why some people have trouble wearing standard, nondigital watches.) This could put mysterious pops and clicks on a tape being handled. Sections of tape might even be partially erased. Try to handle tape only by the edges to minimize such problems. Touch the tape as little as possible. Thin, lint-free gloves can be some help.

To minimize handling of tape, splice long pieces of leader to both ends of the tape. Leader is simply plastic tape with no magnetic coating It is usually clear or a translucent color. It's cheap, and it should always be used. Many recordists feel it is a good idea to regularly use one color leader to indicate the beginning of the tape and a second to mark the end.

You can use a felt-tipped pen on some leader tapes. This allows you to make a note of the tape contents on the leader itself. Be sure to write *only* on the side facing away from the tape heads. Ink from a felt-tipped pen can really gunk up a recorder's heads.

When handling a tape, especially for splicing, be careful not to stretch it. Keep as much tape as possible wound on the reels at all times to prevent contamination.

CLEANING EQUIPMENT AND SUPPLIES

For good performance, all recording equipment *must* be kept scrupulously clean. This is especially true for tape heads. Dirty heads make poor recordings and poor sound on playback. They can even damage the tape. Accumulated dirt can significantly reduce the lifespan of the heads too.

Cleaning the heads is not very difficult. The best way is to use denatured alcohol applied with cotton-tipped swabs. *Do not* use rubbing alcohol. To be safe, it is best to use a solution marketed specifically as tape head cleaner. It is somewhat more expensive, but using the wrong substance could lead to some hefty repair bills.

Usually a cover plate can be lifted off to expose the heads, as illustrated in Fig. 7-1. On most recorders this plate is just held by a pressure joint, although some are held down with a screw. If no screw is visible, gently lift on the plate, rocking slightly back and forth. *Do not force it.* If it doesn't come off easily, and there are no

Fig. 7-1. Usually a cover plate can be removed to expose the heads for cleaning.

directions in the owner's manual, assume it isn't designed to come off. Position the recorder so you can see up under the cover plate. You might need special swabs with long sticks. These are sold in most stores that carry recording accessories. They are illustrated in Fig. 7-2.

Dip one of the swabs into the head cleaning solution. Dab it on each of the heads and any other metallic parts. (Some head cleaning solutions can damage some plastic or rubber parts—be careful). Use each swab for just one or two parts. You don't want to smear any loose dirt from one head to another.

At all times, be very careful not to use too much force. You don't want to bend or break anything. You won't do any damage as long as you work slowly and gently.

Next, take a clean, dry swab and dry off each of the parts you just applied the solution to. Examine the cotton swab carefully. If you can see any dirt on it at all, repeat the procedure with another pair of swabs—one with solution and one dry—until the dry swab doesn't pick up any visible dirt.

Do not assume because the drying swab looks clean that it can be reused. Swabs are cheap. Throw each out after a single use.

When you are finished, put the cover plate back on immediately to prevent dust in the air from settling on the delicate parts.

That's all there is to cleaning heads. It shouldn't take you more than a minute or two. It should be done frequently. If the recording equipment is used an hour or two daily, the heads should be cleaned at least weekly. You can't clean them too frequently. Some professionals automatically clean the tape heads before each use. This might be a little excessive, but it certainly isn't going to hurt anything.

Never try to clean heads with a rag or use any cleaning solution not specifically designed for this purpose. You'll probably do more harm than good.

Cassette recorder heads should be cleaned the same way, although it is usually a little more awkward to get to the heads. Sometimes putting the deck in the play mode with no tape and power disconnected makes the heads more accessible.

You might be tempted to use those head cleaning tapes that are available almost everywhere. They just drop in like a cassette, "play" them for a few seconds to clean the heads, and that's that. These devices are certainly convenient, but beware.

Essentially a head cleaning cassette is like a little rag that is automatically wiped across the heads. What happens to the dirt? It stays on the cleaning tape. The next time you run it across the heads, it may smear old dirt back on, leaving the heads dirtier than before they were "cleaned." Head cleaning cassettes should only be used once and then discarded. At the very most, they should never be used more than three times. Yes, throwing them away after just one or two uses can get pretty expensive very quickly, but you don't save anything if they start doing more harm than good. I cannot recommend these head cleaning cassettes. Cleaning the heads the right way isn't that much trouble. And not cleaning your heads at all is a very bad idea. Do yourself a big favor—do the job right.

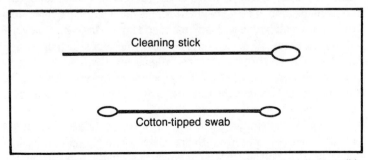

Fig. 7-2. Special cotton swabs with long sticks are used to reach inaccessible heads for cleaning.

DEMAGNETIZATION

You also must beware of "electrical dirt." Anything metallic, and some nonmetallic substances, can pick up a magnetic charge. This includes metal guides in the tape path and the tape heads themselves. This unwanted charge can partially erase high-frequency signals. I have seen one machine with heads so badly magnetized they completely erased any tape as it was played!

Usually the results of head magnetization won't be quite so dramatic. Instead there is a gradually dulling of the high end. Tapes sound muffled and "dirty."

A regular demagnetization procedure is a must for anyone with a tape recorder of nay king. Most people with a tape deck in their stereo system ignore (or don't know about) this important procedure. They are doing irreparable damage to their tapes. Once the high frequencies have been erased by an unwanted magnetic field, there is no way to get them back. No one with a serious interest in recording should neglect regular demagnetization.

Demagnetization is accomplished with a special tool called, naturally enough, a *demagnitizer*. These units can be purchased in almost any electronics/stereo store. Prices range from about $5 to $30. They all do the job well. The differences between models is in convenience features. For example, a demagnetizer with an on/off button is easier to use than one without, but both work just as well as the other.

One feature you should consider is the shape of the probe. Not all demagnetizers will fit in all tape machines. Make sure it is compatible and will reach the heads of your equipment. Some deluxe demagnetizers come with a variety of interchangeable probes to fit almost any tape machine.

The actual process of demagnetization is simple enough. First, move all tapes 6 feet or more from where you will be doing the demagnetization. The demagnetizer puts out a large magnetic field that can wreak havoc with any magnetic tape. Make sure they are all out of range. When in doubt—move the tapes further away.

Now, make sure the tape recorder is off. (Unplugging it may not be a bad idea, even if only to preclude any possibility of it being turned on.) Expose the recorder's heads by removing the cover plate if possible.

Plug in the demagnetizer. If it has no on/off button, hold it out an arm's length away from the recorder while plugging it in. If it does have a button, extend it an arm's length away from the recorder before turning it on.

Next, bring the demagnetizer probe slowly and smoothly towards the tape recorder. Pass it slowly and smoothly over each of the heads and tape guides—anything that comes in contact with the tape that might conceivably be magnetized. Bring the probe as close to the head faces as you can, without actually making physical contact. Be very careful not to scratch the head. Some demagnetizers come with a plastic tip that fits over the end of the probe to protect against scratches. You can also get similar protection by putting a little plastic tape over the end of the probe. Don't use too much and make sure no gummy residue can seep out.

You only have to hold the demagnetizer probe near each part for a second or two. The entire process probably won't take much more than 30 seconds or so.

When finished, extend the demagnetizer again at arm's length from the recorder before turning it off and/or unplugging it. When a demagnetizer is first powered up or powered down, it emits a huge burst of magnetic energy, which could put a permanent magnetic charge on the heads—a charge too strong for the demagnetizer to remove! Always turn it on and off (plug it and unplug it) as far from the recorder's heads as physically possible.

How often should you demagnetize your recorder's heads? It depends on how much the machine is used. Some professional studios demagnetize everything daily. That might be a bit excessive for an amateur or semiprofessional recordist. If you use the equipment daily, once a week should be sufficient. If you only work in your studio on weekends, once every two or three months should do. It is best to be overcautious and demagnetize more often than you need to. If you don't demagnetize often enough, you will do irreparable damage to your tapes, and you may even damage some of your equipment.

For cassette decks (which often have hard to get at heads) special demagnetizers in a cassette housing are available. They usually operate from a miniature battery. These units are a lot more convenient than regular demagnetizers. Just pop them into the deck, run it for a minute or so, and that's that. There is no risk of scratching heads or applying a strong magnetic charge from turning the demagnetizer on or off too close to the heads.

Unlike the head cleaning cassettes, demagnetizer cassettes are a good idea and can be reused. To be on the safe side, always clean the head before demagnetizing. This is especially true for units sold as demagnetizers/head cleaners. Never use the demagnetizer cartridge for cleaning. Once it gets dirty, it will smear dirt around every time you try to demagnetize. Use these units for demagnetization only.

LUBRICATION

Any machinery with moving parts requires some kind of lubrication. This includes tape drives. Most require some relubrication from time to time. You might want to know how to do it yourself. If you aren't sure how to go about this, the best advice I can give you is—if you have to ask, *don't do it*! Improper lubrication can do more harm than good and can even ruin an otherwise perfectly good tape deck. If the manufacturer gives specific information on relubrication in the owner's manual, fine. Follow the manufacturer's instructions to the letter. Don't try any substitutions. Substituting heavy grease for a lightweight oil or vice versa can do a lot of expensive damage.

If the manufacturer does not recommend user lubrication, leave the job to a professional who knows what he is doing. This is one case where doing it yourself is no bargain. The risks are too great.

GENERAL EQUIPMENT MAINTENANCE AND REPAIRS

For best results, your equipment should receive a regular servicing checkup. A technician checks for tiny flaws that might not yet be apparent in use but could cause more extensive damage later, calling for more expensive repairs. The technician also performs any necessary lubrication and internal control tweaking. Never adjust any internal controls yourself, unless you know exactly what you are doing. The technician also cleans the delicate internal mechanism.

A good tape deck under average studio use probably should go in for a regular maintenance check once a year. You might be tempted to let it go to save some money—if it's not broken, why fix it? this is false economy. The odds are good that eventually something will break down due to clogged dirt, loss of lubrication, or something similar. The repair after the damage is done will almost invariably cost more than the regular maintenance check.

If you are a semiprofessional recordist and can't afford to have any of your equipment down for repairs too long, you might want to consider a service contract. It is more or less an insurance policy for your equipment. Many service contracts promise 24-hour repair service. Read the policy carefully and determine if the insurance is worth the price to you. If you are recording as a hobby, you might just want to pay for repairs when they are needed, even if it takes a week or two. Almost all service contracts include regular maintenance checks.

With or without a service contract, always see to it that your equipment gets a regular maintenance check. Recording equipment is too expensive to gamble on.

Chapter 8

Splicing

Anyone who works with magnetic recording tape sooner or later has to make a splice. If you are interested in creative recording, developing a good splicing technique is essential.

Splicing is the mechanical joining of two separate pieces of tape. This certainly sounds simple enough. It's not very hard to make good splices, once you know how to go about it. Unfortunately, it often seems that too few people know how to go about it properly. They make splices any old way. (I've even seen some professionals who were habitually careless about making splices.)

Bad splices are one of the primary sources of grief and frustration in most recording studios. There is really no excuse for this. I suppose it happens because everyone considers splicing so mundane and basic that it isn't worth spending time on it. To this way of thinking, splicing is so obvious that there is no need to teach anyone how to do it. Books on recording techniques might include a brief paragraph or two, but that is not enough. Splicing is a very important skill for the recordist. It should not be shortchanged.

In addition to basic splicing, you might occasionally want to use some nonstandard splices for special effects. These unusual approaches are also explored after covering the basics.

REASONS FOR SPLICING

Splicing is generally done for one of two reasons—editing and repairing tapes. After making a recording, you might want to remove some of the sounds on the tape. As a simple example, let's say you recorded three songs on a reel of tape. One of the musicians made a couple of very audible mistakes during the second song, so you want to delete this song from the tape.

All you have to do is cut the tape at the beginning and at the end of song #2 and discard the central section of the tape. Then it's just a matter of splicing together the end of song #1 and the beginning of song #3. This is illustrated in Fig. 8-1. The end result is the same as if song #2 had never even been recorded in the first place.

Splicing may also be used to change the order of the sounds on the tape. For example, Fig. 8-2 shows a strip of tape with four songs recorded on it in order—1-2-3-4-.

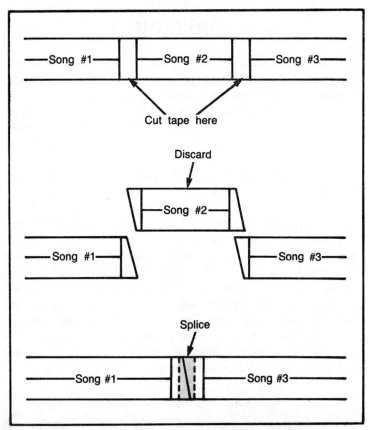

Fig. 8-1. Splicing can be used to delete an unwanted section of a recording.

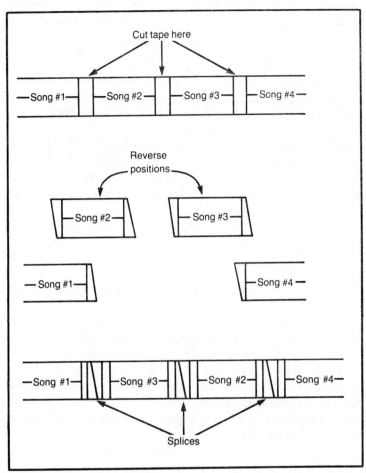

Fig. 8-2. Splicing can also be used to change the order of sounds on the tape.

If we make three cuts—between songs #1 and #2, between #2 and #3, and between #3 and #4—we end up with four pieces of tape. Exchanging the positions of the middle two places before splicing results in a strip of tape with the songs on it in this order—1-3-2-4.

The majority of splices, however, are made to repair broken tapes. If a tape tears or breaks, a splice can reattach the pieces so that the tape can be reused.

MAKING THE BASIC SPLICE

The first step in making a good splice is to make a clean cut at the ends of the tape to be spliced together. A broken tape is likely

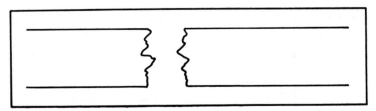

Fig. 8-3. A broken tape is likely to have a rough and ragged edge.

to have a rough and ragged edge (Fig. 8-3). Besides being difficult to align properly, such a ragged edge is quite likely to be very fragile. Small pieces might break off, and the two ends won't match up neatly. This results in a poor, insecure splice. It may be audibly noticeable during playback too.

Before actually splicing the two ends of tape together, you should cut off the ragged ends to get smooth edges to work with. A few fractions of a second of the recorded sound will be lost, but this is rarely noticeable if you cut as close to the break as possible. Even if the lost sound is noticeable, consider the alternative—an unusable tape.

For editing splices, you start out with a smooth, clean cut. There shouldn't be any ragged edges to trim away.

The cut should be angled. Don't make a straight up and down splice, like the one in Fig. 8-4. This would give a minimum amount of surface area across the splice and result in a physically weaker joining. In addition, the abrupt jump from one piece of tape to the other could result in an audible pop during playback, which can be quite objectionable.

A good basic splice is cut at a 45-degree angle (Fig. 8-5). This

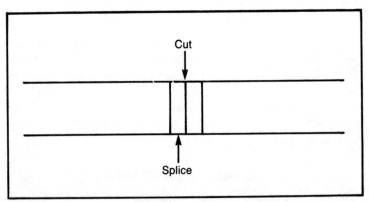

Fig. 8-4. Straight up and down splices are not desirable.

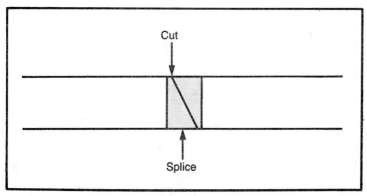

Fig. 8-5. A good, basic splice is cut at a 45-degree angle.

gives the maximum common surface area between the joined pieces of tape for the firmest and most reliable splice. It also provides a gradual change from one piece of tape to the next, minimizing the audible effect.

It is important that the scissors or razor blade used to cut the tape is not magnetized. If there is any magnetic charge on the cutting implement, it can record a loud, permanent pop on the tape. Periodically demagnetize the scissors or razor blade you use for splicing.

The tape itself should be kept very clean. Avoid touching the magnetic coating of the tape as much as possible. Fingerprints and other contamination can cause dropouts and noise. In some cases, such contamination can also interfere with the mechanical integrity of the splice by preventing the splicing tape from sticking as well as it should.

Many recordist wear clean, lint-free cotton gloves when working with their tapes. These gloves are inexpensive enough to be disposable. They are also thin enough to not interfere with dexterity in picking up and handling small bits of tape.

Once a good clean cut has been made, the two pieces of tape are carefully abutted together. They should align perfectly with no gap or overlap between them.

Two typical bad splices are illustrated in Fig. 8-6. In Fig. 8-6(A), the two pieces of tape are not aligned. They won't wind onto the take-up reel neatly and could warp the reel or put a crease in the tape. When this splice passes over the recorder's playback head, the position of the recorded track will shift, resulting in a change of the level of the signal. It doesn't take much of an alignment er-

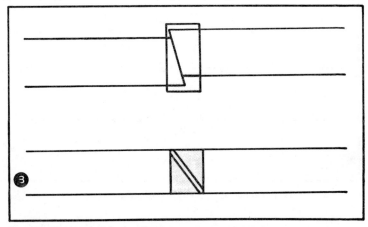

Fig. 8-6. Two typical bad splices.

ror to cause an extreme drop in volume. Crosstalk between adjacent tracks might also increase. On top of all this, the splice is mechanically weak and likely to break apart soon.

In Fig. 8-6(B), there is a gap between the two sections of tape. The adhesive gum on the splicing tape could smear on the head as it passes. Wound on a reel, the splice could stick to the next layer of tape. Also, the gap could result in a clearly audible popping noise when the tape is played back. Again, this bad splice is structurely weak. It is probably just a matter of time before it breaks apart and has to be redone. Why not do it right the first time?

The two pieces of tape must be precisely aligned and abutted exactly with no gaps and no overlap. This can be a little tricky to accomplish until you get the hang of it. With a little practice, it's not too difficult to master. A splicing block (discussed in the next section) can be a big help.

The actual splice is made with special splicing tape. This is a narrow, semitransparent tape. Typical widths are 1/8, 1/4, and 1/2 inch.

Always use specially designed splicing tape. Some hobbyists (and even a few semipros) try to save a few pennies by using regular cellophane tape. This is a bad idea!

The adhesive gum used on standard cellophane tape is simply not designed to withstand the pressures a splice is subjected to. The adhesive gum tends to bleed out from under the splice, making the tape sticky, and can stick to adjacent layers of tape on the reel. Moreover, the gum can adhere to the recorder's heads and drive components, which can cause a lot of damage. *Never use ordinary cellophane tape for splicing!* It cannot do a good job and can do a lot of harm. It is also generally too wide.

The splicing tape is placed gummy side down across the abutted magnetic tape heads on the side facing *away* from the recorder heads. If you put the splice on the side that passes over the heads, you greatly increase the risk of getting adhesive gum on the heads. Also, the added thickness of the splice pulls the tape further from the head's magnetic gap for an instant. This can result in a noticeable drop in the signal's level and clarity.

The splicing tape is placed at a 90-degree angle to the magnetic tape. Do not lay it at a 45-degree angle along the cut.

The splice should be as narrow as possible, but it should be wide enough to cover the entire length of the angled cut, as shown in Fig. 8-7. If the splicing tape doesn't cover the entire cut, the splice will be structurally weak and prone to pull apart.

Too wide a splice should also be avoided. This could distort the magnetic tape as it is wrapped on and off the reels. For general use on 1/4-inch recording tape, use splicing tape with either a 1/4-inch or 3/8-inch width.

Once the splicing tape is placed over the cut, the only remaining step is to trim off the excess splicing tape. You don't want any excess hanging over the edges.

It is a good idea to overtrim slightly, cutting away a sliver of the magnetic tape as shown in Fig. 8-8. This ensures that there won't be any excess width to interfere with the tape winding onto and off of the reels. Don't get carried away and cut too far into

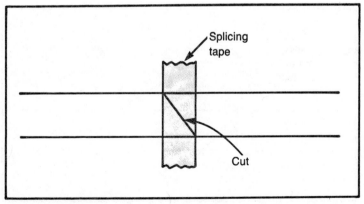

Fig. 8-7. The splice should be as narrow as possible, but wide enough to cover the entire length of the angled cut.

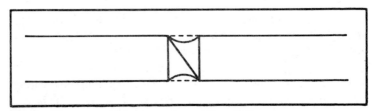

Fig. 8-8. It is a good idea to overtrim a splice slightly.

the magnetic tape. This reduces the track area, and therefore the signal level, and makes a physically weaker splice because there is less material there to hold up to the strain. The trim should be just a barely visible sliver.

Like most skills, making a good splice requires some practice. It is not really difficult, as long as you use reasonable care.

SPLICING BLOCKS

The hardest part of making good splices is getting the two tape edges properly aligned. This part of the job can be done easier by using a splicing block.

A splicing block is not a complicated device. It is simply a firm base with a trough that is exactly the width of the magnetic tape to be spliced. The tape is placed in the trough. If it is not aligned properly, it won't lay flat. The tape naturally tends to settle into the trough indentation. Hold-down arms prevent the tape from moving from the desired position. A typical splicing block is shown in Fig. 8-9. The trough should fit the particular size tape you are using. Most hobbyist and semiprofessional recording work is done

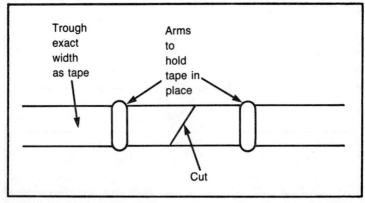

Fig. 8-9. A typical splicing block.

with 1/4-inch tape. A splicing block with a 1/2-inch trough would be virtually useless.

Splicing blocks with a 1/8-inch trough are also available for splicing cassette tapes. Some splicing blocks have dual troughs for 1/4-inch and 1/8-inch tapes.

Usually there is a guideline in the trough to indicate the standard 45-degree cutting angle. Many splicing blocks have a built-in razor handle for cutting the tape. A complete splicer of this type is shown in Fig. 8-10.

The two pieces of tape to be spliced are laid in the trough, overlapping slightly over the cut line. A control on the blade handle is set to "CUT," and the handle is brought down over the two tapes. An internal razor blade cuts the tape neatly at the precise angle. Because both pieces of tape are cut with the same stroke, you can be sure they will fit together neatly.

The handle is then lifted, and the scraps of tape that have been

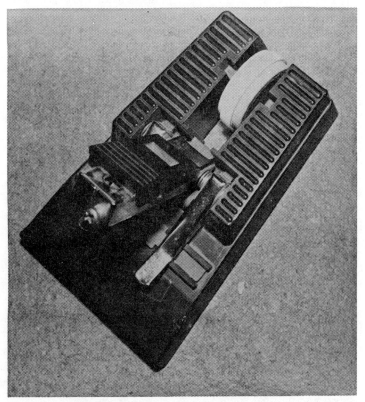

Fig. 8-10. Many splicing blocks have a built-in razor handle for cutting the tape.

cut off are removed. This removal should be done carefully. Avoid moving the pieces of tape to be spliced. The splicer's hold-down arms help keep them in position.

Now a strip of splicing tape is placed across the junction of the two tapes. Many splicers have a built-in holder to keep the splicing tape handy.

The handle's control is now set to "TRIM," and the handle is brought down across the tapes again. This time a different set of internal blades are in cutting position. They neatly trim off the excess splicing tape and a narrow sliver of the magnetic tape on either side. When the handle is raised, the hold-down arms can be released, and the magnetic tape can be lifted out of the trough. You have a perfect, standard splice.

With a good splicer, you should be able to make a good splice in about 10 seconds after just a little practice. Prices start under $5, so there is little reason for any recordist not to own a full splicer, or at least a splicing block. A few tape decks feature built-in splicers, although some of these can be rather awkward to work with.

UNUSUAL SPLICES

Even if you have a complete automatic splicer, it would probably be worth your while to learn to make splices completely by hand. You never know when an emergency might arise—for instance, one of the arms on your splicer might be broken, but you need to make a splice now. Or perhaps you're doing some field recording work, but you forgot to bring along your splicer.

In addition, if you are doing creative work, you might occasionally want to make a nonstandard splice for a special effect. These unusual splices are usually made manually because an automatic splicer is designed strictly for the more common standard splice.

A splicer is still worth having. Even the most avant-garde recordist uses strightforward standard splices at least 90 percent of the time. But occasionally, you might want to try something different.

Incomplete Sounds

Ordinarily, editing splices are made on silences between sounds. Sometimes the creative recordist might want to make a splice within a continuous sound. This can be done to create a number of very novel and unusual sounds.

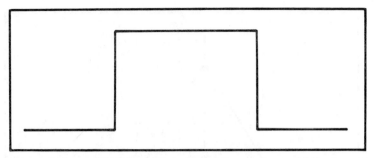

Fig. 8-11. Few real sounds switch instantly between zero and full volume.

One of the most important factors in distinguishing between different sound sources is the amplitude envelope. Few real sounds start at zero, jump up to full volume, then drop instantly back to zero when they're finished, as shown in Fig. 8-11. Instead the instantaneous volume level changes from moment to moment.

Most sounds have a finite *attack* time to build up from zero. Some musical instruments, such as plucked string instruments (e.g., guitar) have a fairly rapid attack. Other sound sources, such as wind instruments (e.g., clarinet) feature a slowly building attack (Fig. 8-12).

Similarly, the time it takes for the sound level to drop back to zero at the end of the sound is also finite. This is called the *decay* or *release*. For most naturally occurring sounds, the decay time is at least as long as the attack time (Fig. 8-13).

Some sounds are held at a more or less constant level for a period of time. This is called *sustain*. Other sounds immediately

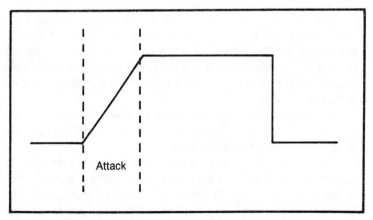

Fig. 8-12. Natural sounds take a finite "attack" time to build up from zero.

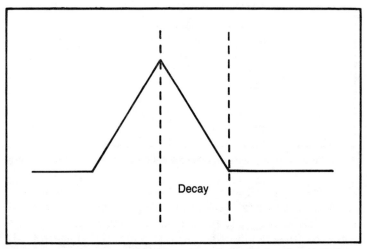

Fig. 8-13. The decay is the time it takes the sound level to drop back down to zero.

begin to fade away (decay) as soon as they reach their peak attack amplitude. Typical envelopes with and without sustain periods are compared in Fig. 8-14.

Most natural envelopes are more complex than this. For example, there may be an initial partial decay before the sustain period (Fig. 8-15). Many natural sounds have a number of partial attacks and decays (Fig. 8-16). Each step of the envelope has a definite effect on the nature of the sound. If we cut out part of the envelope, the sound can be altered drastically. In many cases it can be rendered completely unrecognizable.

Consider the envelope shown in Fig. 8-17A. If you cut and splice the tape at the points shown, you end up with an envelope like the one in Fig. 8-17B. Other effects can be achieved by splicing out the end of the sound (Fig. 8-18). You can also cut out the middle of the sound (the sustain) and splice the attack and decay back together (Fig. 8-19).

Needless to say, these techniques are easiest to work with on relatively long sounds. It would be too hard to cut at the precise desired point for very quick sounds.

To find the desired cutting point, run the tape at the slowest possible speed. It does not have to be the original recording speed. In fact, a slower speed will stretch out the sound, making it easier to find the right spot. Some tape recorders allow you to turn the tape reels by hand with the tape in contact with the heads. You

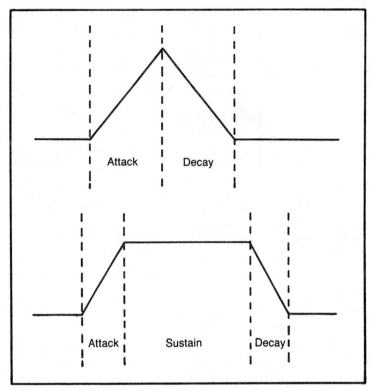

Fig. 8-14. Some envelopes have a sustain period, while others do not.

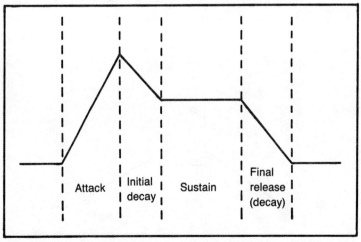

Fig. 8-15. Some sound envelopes have an initial partial decay before the sustain period.

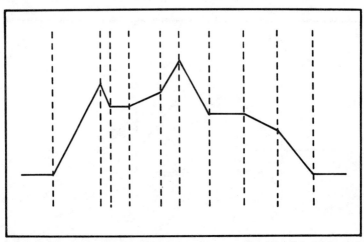

Fig. 8-16. Many natural sound envelopes have a number of partial attacks and decays.

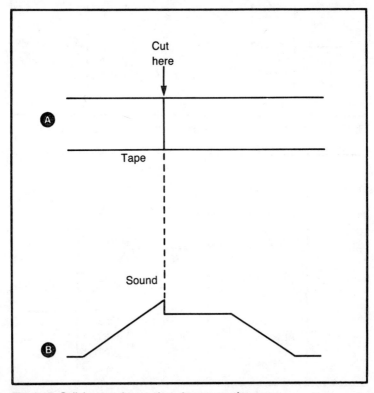

Fig. 8-17. Splicing can be used to change envelopes.

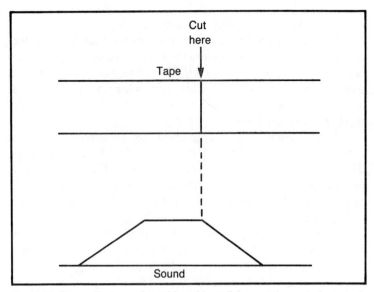

Fig. 8-18. Other effects can be achieved by splicing out the end of a sound.

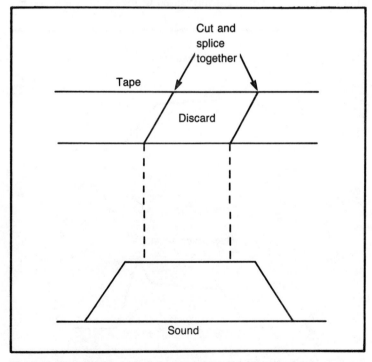

Fig. 8-19. The sustain portion of an envelope can be spliced out.

can move the tape as slowly or rapidly as you like.

When the desired cutting point is positioned precisely over the playback head, mark the back of the tape with a grease pencil. Be careful not to mark the head itself. Now when you remove the tape from the machine, you will know exactly where to cut. This takes some practice so don't get discouraged if you don't get it exactly right the first few times you try it.

This technique is most appropriate for occasional special effects. It could be extremely tedious if you had to do much of it. The early musique concrete composers went through a similar process for each and every individual note or sound in a piece. If nothing else, you certainly have to credit those people for extraordinary patience.)

Another way to manipulate the envelope of a sound is to reverse it, just by running that piece of tape backwards. (This only works for full-track or half-track recorders. The staggering of the tracks on a four-track recorder will defeat your purpose.)

Playing a sound backwards can have a very drastic effect on many sounds. The effect is illustrated in Fig. 8-20. Some sounds

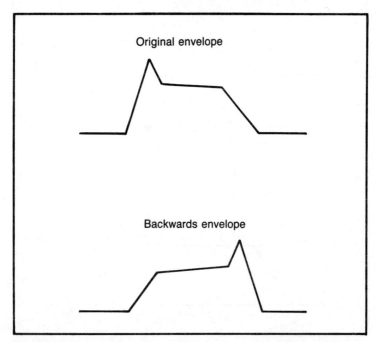

Fig. 8-20. Playing a sound backwards can have a very drastic effect on many sounds.

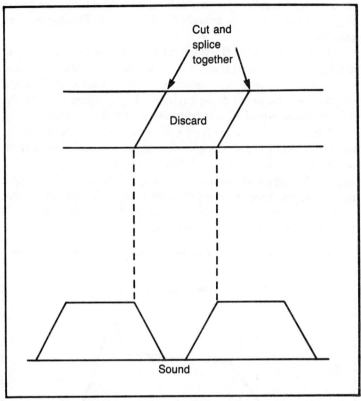

Fig. 8-21. Two partial envelopes can be spliced together to create a new sound.

are rendered completely unrecognizable. Others are only semirecognizable. A backwards recording of a human voice resembles speech in a bizarre, guttural language.

Many backwards sounds sound unnatural because the decay is noticeably shorter than the attack. This almost never occurs in natural sounds.

Creative splicing can also be done to combine two different sounds. For example, you might combine the attack of an oboe with the decay of a banjo. Each of the original sounds is cut in midenvelope, and then the partial sounds are spliced together (Fig. 8-21). A lot of peculiar and fascinating effects can be achieved in this manner.

Unusual Splice Shapes

For certain special effects, you may not want to use a standard 45-degree splice. An unusually shaped splice, like the one in Fig. 8-22, can let one sound blend into the next. This can be very effective.

As less and less of the track area is devoted to sound #1, the amplitude of that sound starts to decrease, while the envelope shape is relatively unchanged. Meanwhile, more and more of the track width is carrying sound #2, so it is increasing in volume. Effectively, it sounds like sound #1 has transformed itself into sound #2.

The *crossover splice* doesn't even have to be straight. For instance, consider the *jigsaw splice* shown in Fig. 8-23. Such techniques are most suitable for full-track recordings where the full width of the tape is devoted to a single signal. With a multiple-track recording, you have to consider the effect on each track used. Also, on a multiple-track recording, each track is quite narrow; it will be very difficult to cut the desired shape across it.

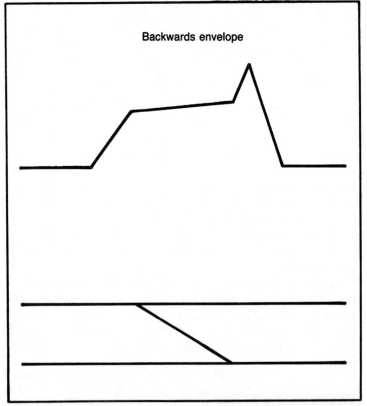

Fig. 8-22. An unusually shaped splice can let one sound blend into the next.

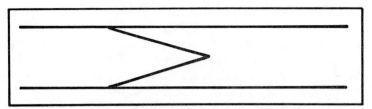

Fig. 8-23. The crossover splice does not need to be straight.

Another serious problem with this technique is that the splices tend to be mechanically weak and much more likely to break apart than a standard splice. This is especially true when very broad cutting angles are used. Still, the effects achieved can be quite intriguing, and it is often worthwhile experimenting with such unusual splicing techniques.

Chapter 9

Mixing

Creative recording is rarely done in one straightforward take. Instead, sounds recorded at different times are combined to make the complete recording. This is called *mixing*.

Many tape decks have limited built-in mixing capabilities (either sound-on-sound or sound-with-sound). More refined mixing is done with two or more tape decks and a mixer.

SOUND-ON-SOUND

Some tape decks feature sound-on-sound capability. Sound A is recorded, then the erase head is disabled, and sound B is recorded on the same track as sound A. If an error is made, both sounds must be rerecorded.

Sound-on-sound is accomplished by disabling the erase head during the second recording pass. If your recorder does not include this feature, you can simulate it by placing some cellophane over the erase head. Do *not* use cellophane tape. Some of the adhesive gum will adhere to the head when the tape is removed. This can ruin tapes later run on the machine.

The cellophane shield does not completely disable the erase head, but it greatly reduces its effect. A more effective, but riskier, approach would be to install a switch to disconnect the erase bias signal, as illustrated in Fig. 9-1. Do not attempt this type of modification if you do not have adequate technical training.

Even with the erase head disabled, some erasing still occurs

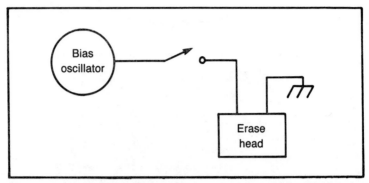

Fig. 9-1. Sound-on-sound can be added to a tape recorder by disabling the erase bias signal.

during the second recording pass, especially at high frequencies. The S/N ratio is also degraded somewhat.

For best results, record the first signal at a higher than normal level to compensate for loss during the rerecording. The first signal should preferably have relatively little high-frequency content. A bass or rhythm line would be the best choice.

Never use sound-on-sound for more than two signals on a single track. The sound will be severely degraded if this technique is overused.

Sound-on-sound is unquestionably a limited mixing technique. Within its limits, it can accomplish a number of useful effects.

SOUND-WITH-SOUND

Sound-with-sound is a more versatile and useful single-deck mixing feature. This feature is only available on three-headed tape decks.

The usual arrangement of the heads in a three-headed deck is shown again in Fig. 9-2. In most straightforward recording, the erase head first blanks out any previously recorded signal on the tape. Next, the tape passes over the record head for the new signal to be imprinted onto the tape. A few milli-seconds after the signal is recorded, the tape passes over the playback head. The recordist can monitor the signal being recorded for over or under modulation and/or distortion. The monitor playback signal is heard briefly after the sound is recorded. There is a short delay between recording and playback. The delay is defined by the distance between the record and playback heads and the speed of tape travel. The smaller the separation between the heads or the faster the tape

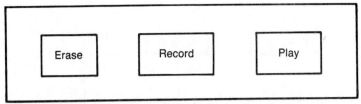

Fig. 9-2. The tape recorder's heads are always arranged in this order.

speed, the shorter this delay will be.

Now see what happens when you try to record a second track in synchronization with the first, previously recorded track. The musician listens to the previously recorded track, usually through headphones (sometimes monitor speakers may be used.) A second part is played along with the first.

To keep things simple, let's look at a single pair of notes. One note is on the originally recorded track. The second note should be simultaneously added on the second track. The musical effect is illustrated in Fig. 9-3.

The musician listens to the first track. As soon as he hears the recorded note, he sounds the new note. The problem is, the old note begins over the playback head, while the new note begins over the record head (Fig. 9-4). The resulting musical effect is shown in Fig. 9-5. This effect can be quite objectionable.

How, then, can sound-with-sound recording be accomplished? The record head for the previously recorded track is temporarily converted into a playback head. This is done with a *sync* switch, as illustrated in Fig. 9-6. Various manufacturers use different names

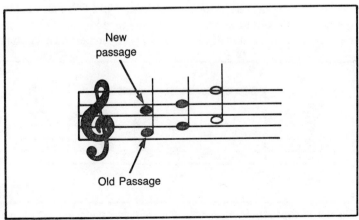

Fig. 9-3. The musical effect desired in the example described.

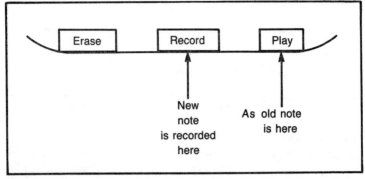

Fig. 9-4. The old note begins over the playback head, while the new note begins over the record head.

for this feature. For example:

- ☐ Multi-Sync (Dokorder)
- ☐ Quadra-Sync (Akai)
- ☐ SEL/REP (Otari)
- ☐ Sel-Sync (Ampex)
- ☐ Simul-Sync (Teac)
- ☐ Synchomonitor (Pioneer)

All of these terms mean pretty much the same thing, and all work in more or less the same way.

The record head is designed for optimum performance for recording. Optimum playback has somewhat different requirements. Consequently, when the record head is used for playback in the sync mode, the frequency response and S/N ratio suffer noticeably. Because the purpose of the sync mode is synchronization of tracks, the sound quality is not of prime importance. As long

Fig. 9-5. The result of trying to dub the musical passage of Fig. 9-3 without sound-with-sound capabilities.

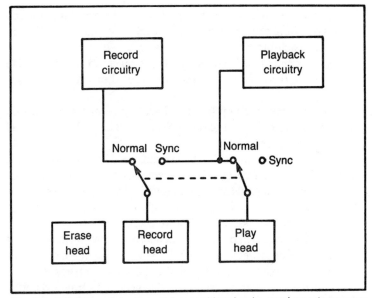

Fig. 9-6. A SYNC switch allows the record head to be used as a temporary playback head.

as you can hear what is on the first track with reasonable clarity, everything is fine. You are interested in timing considerations, not high-fidelity playback.

If your tape deck does not have a sync switch, a competent technician might be able to add one. I would not recommend that you try to install one yourself, unless you know your way around an electronic circuit. It's not hard if you know what you're doing. The hardest part is usually finding a good place to install the switch on the recorder's control panel.

FOUR-CHANNEL DECKS

A 4-track recorder deck is very well suited for mixing. If you can afford an 8-track or 16-track recorder, you'll have even more versatility. But four tracks is sufficient for most applications, although a few more steps may be required.

Multitrack recording is used in virtually all professional recordings. Up to 32 tracks are recorded either simultaneously with separate directional microphones or at different times using a combination of the two. Once the recording is completed and properly balanced, it is mixed down to two-channel stereo.

Virtually all 4-track tape decks have a synchronization feature.

It's easy to see how four separate parts can be recorded on a 4-track recorder. Simply record one part on each track, keeping them in step with each other via the sync feature.

By combining sound-on-sound techniques, up to 10 separate parts can be recorded on a 4-track recorder. Let's see how this is done.

Part 1 is recorded on track 1. Part 2 is recorded on track 2, while monitoring the first part (SYNC). Part 3 is similarly recorded on the third track.

So far nothing unusual has been done; but when the fourth track is recorded, the procedure changes somewhat. The outputs from the first three channels are mixed together along with part 4. All four parts are recorded monophonically on track 4. Tracks 1, 2, and 3 are now erased and can be reused.

Part 5 is recorded on track 1, synchronized with track 4. Part 6 is similarly recorded onto track 2. Now the signals from tracks 1 and 2 are mixed with part 7 and recorded onto track 3.

Part 8 is recorded on track 1 and synced with tracks 3 and 4. Track 1 and part 9 are mixed together and recorded onto track 2. Finally part 10 is recorded on track 1, in sync with the other three tracks.

Now 10 independent parts are recorded on four channels. The arrangement of parts is summarized in Fig. 9-7.

This technique, which is sometimes called *track bouncing*, can be extended for even more parts, but it is best to limit yourself to

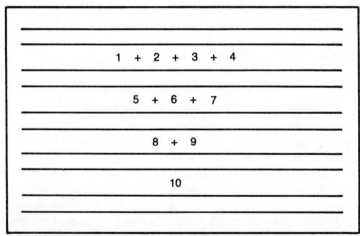

Fig. 9-7. Up to 10 parts can be recorded on a 4-channel tape deck using sound-with-sound techniques.

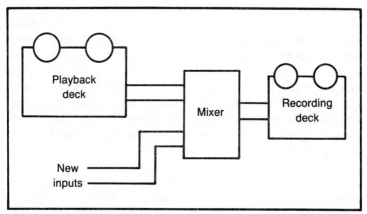

Fig. 9-8. Four-track capability can be simulated with two 2-track recorders.

10 parts on four tracks. Each time a track is rerecorded, the noise and distortion increase. If you intend to later mix the tape down to 2-channel stereo, some of the tracks will have been rerecorded twice. It's probably not a good idea to rerecord a track more than twice without noise reduction equipment (see Chapter 10). Obviously an 8-track or 16-track recorder allows for even more parts.

It is important to be very careful in the mixing stages. If the balance (relative volumes) of the various parts is not right, or a mistake is made in the new part, the mix has to be done again. Don't erase the old tracks until you are 100 percent positive that the new mix-down track is what you want. Once the parts are mixed together, there is no way to separate them again. To change any of them, you would have to redo them all.

THE MULTIDECK APPROACH

Suppose you don't have access to a 4-channel tape deck. What if you need more versatility than a single deck can handle? Are you out of luck? Not at all!

Four-track capability can be simulated with two 2-track recorders connected in a mixing configuration similar to the one shown in Fig. 9-8. The first deck is used to play back previously recorded tracks. The second track records the mixed new tracks.

A pair of 4-track decks connected this way gives simulated 8-track capability. One 4-channel deck and one 2-channel deck add together for a simulated six tracks. The multiple deck approach also gives you many capabilities that are not available with a single deck with the same number of tracks.

The speed of the playback deck can be varied. For example, let's say you need a very rapidly played passage, and you just aren't dexterous enough to carry it off at the required speed. Transpose the passage down one octave and play it at half the desired speed, recording it at 3 3/4 ips. Now play it back at 7 1/2 ips and mix it with additional parts at the regular tempo. Doubling the tape speed doubles the speed of the passage and doubles all frequencies (raises all notes one octave). You should be aware that this trick alters the harmonic relationships in the sounds, altering the sound quality somewhat.

Some tape decks feature a dynamic (continuous) speed control which can be varied while the machine is running. This allows for a number of special effects. Speeding up or slowing down a sound can change its nature considerably. It is usually awkward to manipulate the speed control during the recording phase, especially if you are also the musician. Moreover, if you don't get the speed changes quite right, and it generally requires a few takes to get it right, the musician will have to play the passage each time you try. It is more convenient to vary the speed during playback, dubbing the altered sound onto a second, constant speed recorder.

Another advantage of the multideck approach is that it is easier to insert a noise reduction device than with a single recorder. You can also save the original track tapes as long as you want. If you decide to remix later, it is no problem. All of the tracks do not have to be rerecorded from scratch. The multideck approach to dubbing also gives you more control of the locations of the various sounds (which final stereo tracks they will appear on).

Let's say we are recording a trio for flute, guitar, and piano. We want the stereo image to appear as shown in Fig. 9-9. The flute should be on the right track, and the guitar should be on the left track. The piano is in the middle; that means, it should be on both stereo tracks. This would be difficult to accomplish bouncing tracks with a single deck. With two decks connected through a mixer, it would be no problem at all. The mixer determines how much of each input signal (the previously recorded tracks or the new material) will go to each of the outputs (the tracks to be recorded).

Even more versatility can be achieved by adding a third deck to the system (Fig. 9-10). The new deck is also used for playback. Two previously recorded tapes can be mixed together.

Synchronization between the two playback decks can be a problem. This approach works best when one of the previously recorded tapes does not need to be tightly synchronized. For instance, this

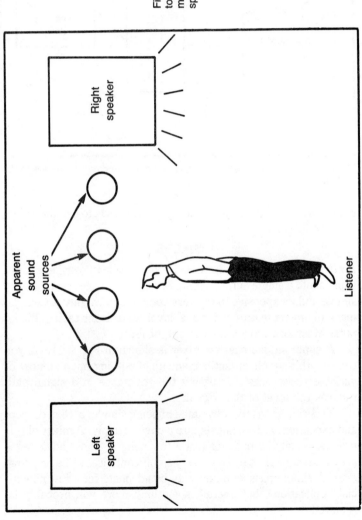

Fig. 9-9. The stereo image can seem to "position" the various instruments between the playback speakers.

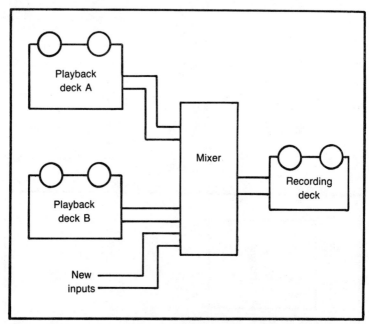

Fig. 9-10. Even more versatility can be achieved by adding a third deck to the system.

system works very nicely for adding sound effects to a radio drama.

MIXERS

A *mixer* is an electronic circuit used to combine two or more input signals into a single output signal. Mixers can range from the very simple and inexpensive to the very complex and highly priced. Almost all mixers feature individual level controls for each of the inputs. Most also have a master output level.

A super simple passive mixer is shown in Fig. 9-11. As you can see, this simple circuit is made up of nothing but a number of variable resistances. Increasing the resistance in a signal path reduces the level of that signal.

This circuit has the advantage of being simple, both in concept and execution. It does not require a power supply. A mixer of this type can easily be built for just a few dollars. These advantages are more than offset by some very significant disadvantages, however. A simple passive mixer may be adequate for a few noncrucial applications, but overall, something more will probably be required.

What are the disadvantages of the passive mixer? For one thing, since it is a passive rather than an active device, it "steals" its operating power from the signal it is passing. The output level is always less than the input level. Since the signals you work with in the studio (especially microphone outputs) are often fairly small, the loss might be more than you can accept. The S/N ratio will often be significantly degraded.

Another important problem with passive mixers is the almost inevitable impedance mismatch between the inputs and outputs. Remember that impedance is made up of dc resistive and ac reactive components. The passive mixer functions by altering the dc resistance in the signal path. Thus, the impedance is changed each time one of the control settings is altered. The odds of the desired level settings giving a good impedance match are very slim to virtually nonexistent.

Better mixers for serious recording work are more complex. They include active circuitry to provide preamplification and impedance matching. Good mixers have an independent preamp stages for each input line and a master preamplifier for the output. The basic mixer has two or more inputs and a single output. Some deluxe models have multiple, independent outputs—usually two for stereo.

Each of the individual input channels might be directed to either or both of the outputs. Usually a panning control, called a *pan-*

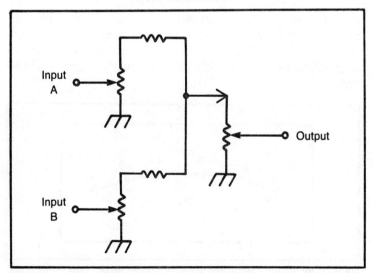

Fig. 9-11. A simple passive mixer can be built with a few variable resistances.

pot, is provided for each of the inputs. The panpot allows the signal to be "positioned" between the outputs. At one extreme setting, the signal is applied entirely to the right output. At the other extreme setting, the signal is applied entirely to the left output. At in-between settings of the panpot, the signal is applied to both of the outputs. The relative levels for each channel are determined by the setting of the panpot. At its midpoint, the signal is split equally between the two outputs.

Some professional mixers have more than two outputs. Multiple panning controls might or might not be provided.

Various equalization filters might be included in the mixer. Provisions for reverberation (artificial echo) effects may also be included. These and other special effects are discussed in Chapter 10.

Professional and semiprofessional mixers have level meters (usually VU meters—see Chapter 3). A typical set of mixer meters are shown in Fig. 9-12. These meters are very useful in allowing you to precisely control the levels of the various signals being mixed and the total output signal.

A typical professional mixer is shown in Fig. 9-13. The rear panel (Fig. 9-14) has a series of jacks for connecting the various inputs and outputs to the mixer.

Whatever its features, a mixer offers considerable power and convenience to the creative recordist.

OVERDUBBING

Later remixing might occasionally be required. For example, new material might need to be added to the previously mixed recording. This is called *overdubbing*. Narration might be added over music, for instance, or music might be added under dialogue for a radio drama.

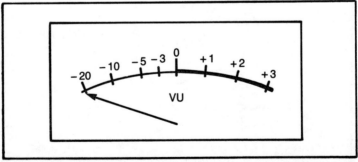

Fig. 9-12. Professional and semiprofessional mixers usually have level meters.

Fig. 9-13. This is a typical professional type mixer.

Overdubbing is not really different from the regular types of mixing discussed throughout this chapter. The primary difference is in the function. Also, overdubbing rarely requires precise synchronization. This makes using two playback decks together with one recording deck more convenient.

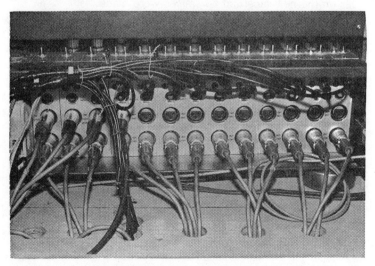

Fig. 9-14. The rear panel of a professional mixer has a number of jacks for connecting the various inputs and outputs in the mixer.

EDITING IN THE MIX

You can do a lot of your editing in the mixing room, eliminating the need for many splices. For instance, consider one of the examples discussed in Chapter 8. Four songs are recorded on the original tape in this order:

1-2-3-4

You want the final tape to have this order:

1-3-2-4

Chapter 8 described how this could be accomplished using splices to physically rearrange pieces of the tape. In some cases, it might be more convenient to do this in the mix.

Put the original tape on the playback deck and a blank tape on the record deck. Play song #1 on the playback deck and rerecord it on the second tape deck. Then stop the recording deck and fast-forward the playback deck to the beginning of the third song. Play song #3 and rerecord on the second deck. Stop the recording deck again, but rewind the playback deck to the beginning of song #2. Rerecord this song. Finally, fast-forward the playback deck past song #3 to rerecord song #4. The songs are now in the correct order on the new tape and no splices were made.

In some applications this method might be considerably more convenient than physically splicing the tape. In other applications, it might be easier to go ahead and splice the tape directly. Both work, but sometimes one method is handier than the other. In deciding which method to use, rely on your common sense and personal preference.

Chapter 10

Special Effects and Accessories

Now that you have mastered the basics of recording, you are ready for the "frills" or many special effects that are possible. This chapter discusses a number of accessories that can come in handy in the recording studio.

EQUALIZATION

Equalization is the process of boosting or cutting the levels of selected frequencies to get something closer to flat frequency response from the system, to "sweeten" the sound to taste, and for special effects.

Basic equalization is not a frill. Some degree of equalization capability is virtually a must for any serious recording setup. It is included in this chapter because equalization devices are usually in external accessory units. Some mixers, preamplifiers, and amplifiers might include some limited provisions for frequency equalization.

Filtering

Before discussing equalizers, you must understand filters. A *filter* is a frequency-sensitive circuit. It passes some frequencies, but blocks others. There are four basic types of filters that respond to different frequencies in various ways. These are:

- ☐ Low pass
- ☐ High pass
- ☐ Band pass
- ☐ Band reject (or notch)

Each of these names are more or less self-explanatory. A *low-pass filter*, for example, passes low frequencies but blocks high frequencies. A frequency response graph for a theoretically ideal low-pass filter is shown in Fig. 10-1. The dividing point between the passed and blocked ranges is called the *cutoff frequency*.

A practical filter circuit cannot instantly switch from totally passed to totally blocked. Instead, there is an intermediate region which the frequencies are partially attenuated. A frequency response for a typical, practical low-pass filter is shown in Fig. 10-2.

The steeper the cutoff slope, the better the filter, and the more distinct its action will be. The cutoff slope is measured in terms of × dB per octave. For example, a filter with a 6 dB per octave cutoff slope and a cutoff frequency of of 300 Hz would add 6 dB of attenuation each time the frequency is raised by one octave (doubled):

```
300 Hz        0 dB  (reference)
600 Hz       -6 dB
```

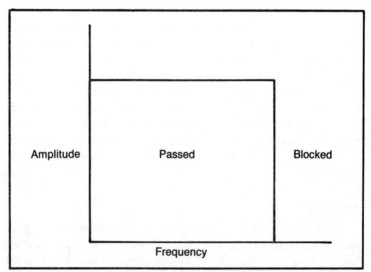

Fig. 10-1. A low-pass filter blocks high frequencies, but allows low frequencies to pass on through to the output.

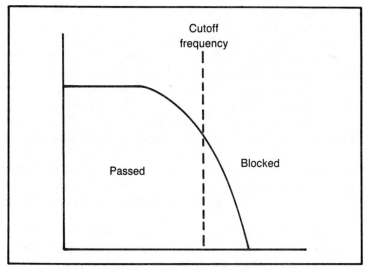

Fig. 10-2. A practical low-pass filter has a finite slope between the block and pass regions of its response.

1200 Hz	−12 dB
2400 Hz	−18 dB
4800 Hz	−24 dB
9600 Hz	−30 dB

The simplest type of low-pass filter is a *RC* (resistor-*capacitor*) network (Fig. 10-3). High frequencies are shunted to ground through the capacitor to ground. Low frequencies cannot pass through a capacitor; therefore, the low-frequency components are passed on to the output.

This simple passive circuit has a lot of major shortcomings. Its operating power is stolen from the input signal. All frequencies are attenuated to some degree because of the resistance in the signal path. High frequencies are not completely blocked, although they

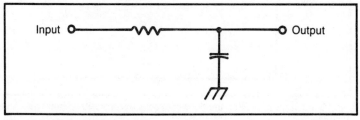

Fig. 10-3. A simple RC network can serve as a passive low-pass filter.

can be highly attenuated. Impedance matching can be tricky. It isn't easy to change the cutoff frequency. The S/N ratio can be degraded because the entire signal is significantly attenuated.

The cutoff slope of a passive filter is very shallow—only about 3 dB per octave. The passive filter is only presented here because it is so simple conceptually; it makes a good illustration of how filters work.

Practical filter circuits for serious work include active components. Basically they are frequency-sensitive preamplifiers. The signal is boosted (amplified) to compensate for insertion loss. Impedance can be matched very precisely. The cutoff frequency can be easily changed with a straightforward front panel control. Most active filters typically have a cutoff slope of 12 or 24 dB per octave.

Modern filter circuits usually employ *op amp* (operational amplifier) circuitry. This results in relatively inexpensive and compact circuits with high efficiency and performance. Any op amp generates some noise, however. Multistage units, especially those using inexpensive op amps, can seriously degrade the S/N ratio.

A *high-pass filter* is just the opposite of a low-pass filter. High frequencies are passed and low frequencies are blocked, as illustrated in Fig. 10-4. In fact a passive high-pass filter can be created by reversing the position of the components of a low-pass filter.

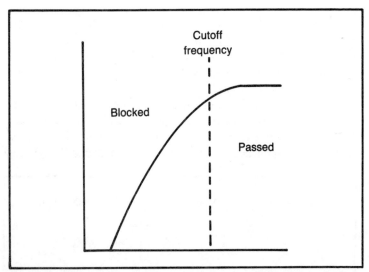

Fig. 10-4. A high-pass filter functions in exactly the opposite manner as a low-pass filter.

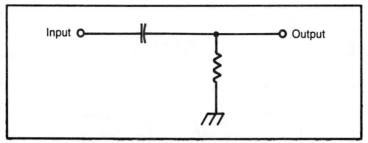

Fig. 10-5. Reversing the positions of the resistor and capacitor can convert a passive low-pass filter into a passive high-pass filter.

The passive circuit is shown in Fig. 10-5.

A more complex type of filter is the *band-pass filter*. Frequencies within a specific band are passed. Any frequency component outside the defined band is blocked. Frequency response graphs for two typical band-pass filters are shown in Fig. 10-6. Notice that the passed band can be either very wide or very narrow or anywhere in between. This is often referred to as the Q of the filter.

Notice that instead of a cutoff frequency (actually there are two cutoff frequencies—one at either end of the passed band), the center frequency is specified. The *center frequency* is the midpoint of the passed band.

The opposite of a band-pass filter is the *band-reject filter*. All frequency components are passed except for those within the specified reject band, as illustrated in the frequency response graph of Fig. 10-7. Because of the appearance of the frequency response graph, this type of filter is often called a *notch filter*.

All of these types of filters are used for equalization. Band-pass filters are generally the most important and versatile.

Tone Controls

Many pieces of consumer audio equipment feature simple equalizers or *tone controls*. Usually there is a bass control (low-pass filter) and a treble control (high-pass filter). Each of these controls can either attenuate or boost a broad band of frequencies. Some devices also feature a midrange (band-pass) control. Sometimes this is called a *presence control*.

These controls allow gross adjustment of the system's overall frequency response. Because each of these controls affects a broad band of frequencies, they cannot really fine-tune the frequency response.

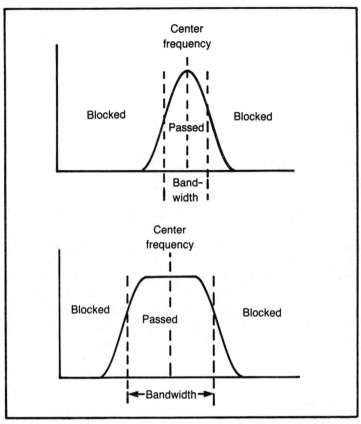

Fig. 10-6. Band-pass filters can have a number of different types of frequency response.

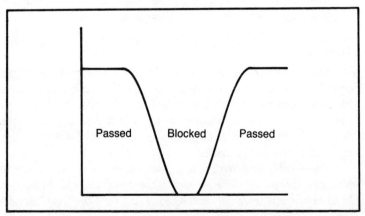

Fig. 10-7. The opposite of a band-pass filter is a band-reject (or notch) filter.

For example, if your system has a peak at 11,000 Hz and a dip at 10,000 Hz, there is no way the standard tone controls can deal with both problems simultaneously. Besides, boosting the 10 kHz dip also boosts the frequencies that are close to flat, giving them unwanted peaking.

Tone controls can be useful on a home stereo unit for adjusting the overall sound to suit the listener's individual taste. They are of little use in the recording studio, however.

Graphic Equalizers

Graphic equalizers are becoming increasingly popular in home stereo systems. They offer more control over the sound than the simple tone controls described in the preceding section.

Basically the graphic equalizer is a series of band-pass filters. Each controlled band can be either attenuated (cut) or amplified (boosted). The number of bands varies.

The simplest graphic equalizers have five bands:

- ☐ Low bass
- ☐ Midbass
- ☐ Midrange
- ☐ Midtreble
- ☐ High treble

Each of these bands covers approximately a two-octave range. Such a device does not allow very fine distinction between adjacent frequencies, but it is a major step up from simple tone controls. The listener can custom-tailor the overall response characteristics to suit individual tastes.

Most graphic equalizers have 10 control bands—each covering a 1-octave range. This allows more detailed control of the system's frequency response. A 10-band equalizer should be sufficient for most home listening conditions.

Deluxe graphic equalizers divide the audible range into 1/3-octave bands and have 27 to 30 independent controls. This allows very precise adjustment of the system's frequency response. A graphic equalizer with more than 30 controls probably would not be very practical. The benefits would not offset the added inconvenience and would get into the "law of diminishing returns."

Graphic equalizers use sliding *potentiometers* for the controls (Fig. 10-8). With sliding controls, the operator can see the control

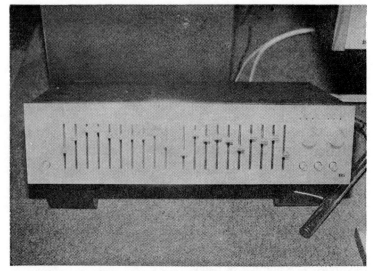

Fig. 10-8. Graphic equalizers feature sliding potentiometers as controls.

settings at a glance. The positions of the various controls "draw" a simple graph of the equalizer's effect on the system frequency response, as shown in Fig. 10-9.

The center position, often provided with a detent to locate it easily, is the flat response level. The input signal in the controlled frequency band is neither amplified nor attenuated. The output in this band is (theoretically) identical to the input.

If the control is raised above the flat (0 dB) position, the controlled band is boosted (amplified). Lowering the control position below the flat point, cuts (attenuates) the amplitude of the controlled frequency band.

The control is generally calibrated in dB. The flat midpoint is the reference 0 dB level. Boost positions are given positive dB values. Cut positions are marked with negative dB values.

A graphic equalizer can be a very valuable component in any good stereo system. It can also be used in the recording studio to good advantage. A graphic equalizer can be used to either compensate for frequency response shortcomings in the equipment or the studio itself or to create special effects by creating deliberate (controlled) deviations from flat frequency response.

Console Equalizers

In a home stereo system an equalizer can only affect the entire signal within the controlled frequency band. If two instruments produce the same frequency, both will be simultaneously affected by the equalizer. For listening to prerecorded material, this is fine. The recordist needs greater control, however, ideally full control over each sound being recorded.

Recordings are generally made with multiple inputs. Each microphone line (or whatever) should have its own equalization controls. Because many recording sessions involve four or more microphones at the same time, a full graphic equalizer for each input is quite impractical. It would be horribly expensive, and all those controls would take up a lot of room on the control board. The engineer might not be able to easily reach all of the controls from a single position.

Many recording studios use a somewhat stripped-down type of equalizer called a *console equalizer*. Interestingly, in some ways a console equalizer can be more powerful than a graphic equalizer.

A console equalizer generally has only three bands—bass, midrange, and treble. The cutoff frequency for each of the bands is switch selectable. For example, a typical console equalizer might offer the following control frequencies:

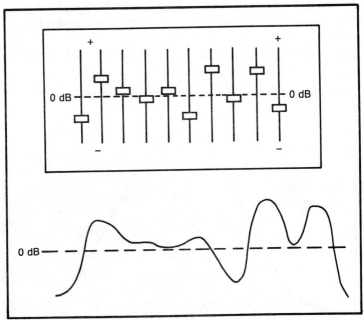

Fig. 10-9. The graphic equalizer's controls "draw a graph" of the equalizer's filtering action.

BASS	MIDRANGE	TREBLE
50 Hz	200 Hz	2000 Hz
80 Hz	320 Hz	3200 Hz
120 Hz	400 Hz	5000 Hz
200 Hz	620 Hz	8000 Hz
320 Hz	1000 Hz	12000 Hz
	1500 Hz	
	2000 Hz	
	3200 Hz	

Notice that there is some overlap between the frequency ranges. Only one frequency per range can be controlled at any one time.

The filtering shape for the bass and treble sections can often be selected with a two-position switch. The two available filter shapes are usually labelled "Peak" (Fig. 10-10) and "Shelf" (Fig. 10-11). The midrange section is strictly of the PEAK type.

A second control for each frequency range adjusts the amount of boost or cut for the selected frequency. This control is usually calibrated in dB.

An independent console equalizer for each input channel can fit easily on a mixer board. A typical control arrangement for this type of equalizer is shown in Fig. 10-12. There is also no law that says console equalizers can't be supplemented with a graphic equalizer if more control is needed.

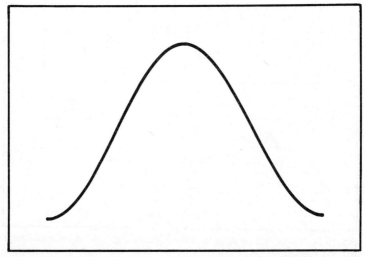

Fig. 10-10. Console equalizers may be set up for PEAK filtering.

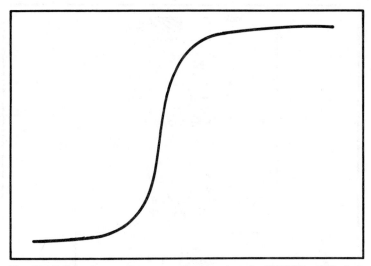

Fig. 10-11. Console equalizers may also be set up for SHELF-type filtering.

Parametric Equalizers

The most powerful and versatile type of equalizer is the parametric equalizer. As the name suggests, each parameter of the filter action is independently variable.

What are these parameters? The parametric equalizer is essentially a band-pass filter, so there are three basic parameters of interest:

- ☐ Center frequency
- ☐ Bandwidth (or Q)
- ☐ Amount of boost/cut

Usually, potentiometers are used to control all three major parameters, rather than discrete stepping switches. Each of the values are continuously variable.

The center frequency control needs no special explanation. It is simply the midpoint of the frequency band to be controlled. For more information, see the description of band-pass filters earlier in this chapter.

The controllable bandwidth (Q) is the most unique and powerful feature of the parametric equalizer. On most other types of equalizers, the bandwidth is fixed. The range of frequencies controlled is not adjustable.

A narrow bandwidth affects only a small range of frequencies. This is used to remove unwanted ringing or buzzing tones without

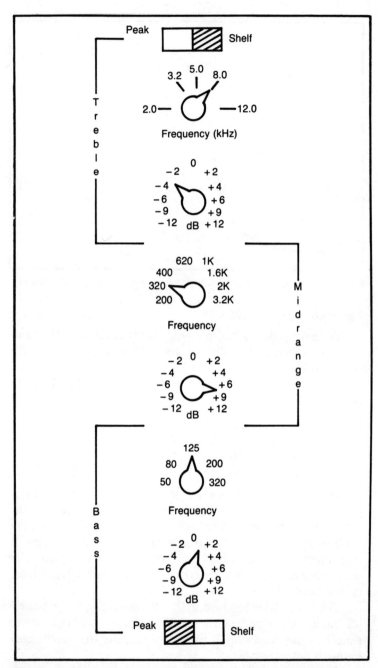

Fig. 10-12. A console equalizer's controls don't take up much space on the mixer board.

severely affecting the overall tonal characteristics of the sound. Boosting a narrow bandwidth of frequencies can help "spotlight" a tone, providing a purer sound. This can be used to clean up "muddy" sound.

Narrow bandwidth equalization should not be overused. On a different set of speakers, some unpleasant resonances might show up.

Wide bandwidths function similarly to console (peak-mode) or graphic equalizers. Because all controls are continuously variable, repeatability can sometimes be a problem with parametric equalizers.

Spectrum Analyzers

Equalization can be done by ear—adjust the controls until things sound good. In some cases, more exact control may be desirable. The way to do this is to measure the strength of each frequency with a spectrum analyzer.

A microphone, with as flat a response as possible, "listens" to the sound in the room. A series of band-pass filters breaks up the total sound into a number of frequency bands. The energy in each band is then displayed.

A spectrum analyzer indicates the actual frequency response of the complete system, including the room effects. This allows very exact equalization to achieve flat frequency response.

NOISE REDUCTION

One of the most significant factors working against high-fidelity sound reproduction is noise. Any piece of electronic equipment adds some degree of noise (extraneous signal) to the input signal. More significant for your purposes is the noise inherent in the tape medium itself.

When a tape is recorded, some of the magnetic particles are randomly aligned. Along with the signal you wanted to record on the tape will be a continuous hissing sound. The hiss has a more or less constant level, regardless of the amplitude of the recorded signal. For loud sounds, the background hiss is drowned out. When the recorded signal is at a very low volume, the tape hiss is quite audible. In fact, for very weak signals, the desired signal might even be drowned out by the noise.

A number of noise reduction devices are now available to help reduce this problem. The noise reduction units for the consumer market (as opposed to expensive professional equipment) include two types: single-ended and dual-ended.

A *single-ended noise reduction system* works only during playback. Nothing special has to be done during recording. For that reason, a discussion of such systems is not really within the scope of this book.

Dual-ended noise reduction systems, on the other hand, involve encoding during recording and decoding during playback. From now on in this chapter, when I say *NR*, I am referring to a dual-ended noise reduction system.

Most practical NRs in use today are variations on the compander concept. *Compander* is a contraction of "COMpresser" and "exPANDER."

Any recording media has a limited dynamic range. If the signal is too strong, the tape will be overloaded, and severe distortion will result. If the signal is too weak, it will be overpowered by the noise floor of the tape. This is illustrated in Fig. 10-13.

A compressor, as the name suggests, compresses the dynamic level of the signal. Weak signals are boosted, strong signals are attenuated (Fig. 10-14).

Loud sounds are reduced in level to prevent the tape from saturation distortion. Soft sounds are increased in amplitude to raise them a comfortable level over the noise floor, which is more or less constant. The compression works only on the signal to be recorded. The level of the noise generated by the tape and electronics of the recorder is not affected because it is generated after the compressor. The recorded signal is represented in Fig. 10-15.

On playback, a complementary expander is used to restore the

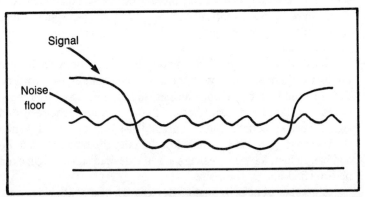

Fig. 10-13. If the recorded signal is too weak, it can be overpowered by the noise floor of the tape.

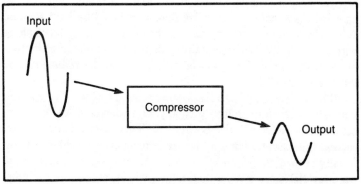

Fig. 10-14. A compressor boosts weak signals and attenuates strong signals.

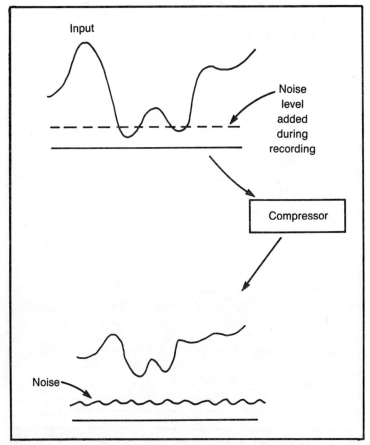

Fig. 10-15. Tape noise is not affected by the compression process during recording.

original dynamic range. Loud signals are made louder, and soft signals are made softer. The noise floor is masked by loud sounds. When a weak sound is being played, the noise level is reduced along with the desired signal, maintaining their relative relationship; that is, even the weak signal will be stronger than the reduced noise level (Fig. 10-16).

The most widely used NR system around today is *Dolby B*, developed by Dr. Ray Dolby. Virtually all modern high-fidelity cassette decks include Dolby B circuitry. This system is created by raising the cassette from a low-grade (primarily voice) to a high-fidelity recording system. The low speed of cassettes makes the need for NR more critical than higher speed reel-to-reel systems. (A reel-to-reel recording can certainly benefit from NR, but it is possible to make good recordings without it.)

Because noise is more of a problem at high frequencies, the Dolby B system functions only above about 5000 Hz. By ignoring the lower frequencies, circuit complexity and cost are reduced. At

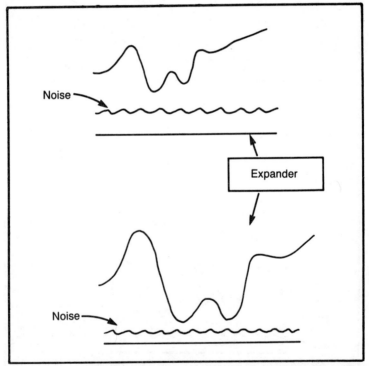

Fig. 10-16. On playback, a complimentary expander reverses the action of the prerecording compressor. Noise is reduced as a weak signal to a lower level.

the time Dolby B was introduced, a full range NR circuit was not reasonably affordable for home stereo systems. Dolby B improves the S/N ratio by about 10 dB at frequencies above approximately 5000 Hz.

During recording, the Dolby circuitry detects the average amplitude of the high-frequency signal to be recorded. The recording level is progressively increased for low-amplitude, high-frequency signals. Thus, the high-frequency signals are recorded at a stronger level than the noise floor.

When the tape is played back, the Dolby circuit performs in reverse. Theoretical flat frequency response is restored to the recorded signal. When the high-frequency signal level is reduced, so is the noise. The S/N ratio is the same as was recorded. Tape hiss and other high frequency noise is significantly reduced, although not completely removed, while the recorded signal is (theoretically) unaffected by the process.

A Dolby B encoded tape can be played back without decoding, although it sounds rather overbright. Turning down the table control usually makes the tape listenable.

While Dolby B does offer some advantages, perhaps its time has passed. Noise in a Dolby B tape is reduced, but it is still audible. Better, more efficient NRs have been developed in recent years. (Some will be discussed shortly.)

Dolby B is designed only to cut down high-frequency noise. While tape hiss is the most objectionable type of noise, it is not the only type. Low and midfrequency noise are completely ignored by this system. I've found that many Dolby B encoded tapes played back without decoding have even more hiss than a comparable unencoded tape.

Some listeners find that Dolby B encoding and decoding colors the sound somewhat. Most listeners don't seem to hear this effect. There doesn't seem to be any correlation between training and the ability to hear this Dolby B coloration—either you hear it or you don't. I find the rather "metallic" sound quality of a Dolby B encoded tape to often be more objectionable than the noise (assuming a well-made recording).

There are still thousands of Dolby B encoded tapes around. Despite its inadequacies in the face of more recent developments, Dolby B is likely to be around for a long time, at least in cassette decks. Any NR system that hopes to usurp the Dolby throne had better be at least partially compatible with Dolby B recordings.

Dolby B circuitry is included in the majority of stereo cassette decks made today. Whether you want it or not, you pay for the components and the licensing royalties. If you find Dolby B adequate for your home system, great. There's no sense in buying another system if the standard one is sufficient for your needs.

I definitely do not recommend Dolby B for your recording studio. For one thing, you should be using reel-to-reel rather than cassette equipment. While Dolby B is included on a few reel-to-reel decks, generally it requires an outboard unit. Considering modern developments, it's not worth the price. Better NR systems are available.

For professional recording, Dolby A is often used. This system is too complex and expensive to put on the consumer market.

Dolby has recently developed an improved version of their NR system. This is Dolby C. Many commercial cassette decks now include circuitry for Dolby C in addition to Dolby B. Dolby C is essentially compatible with Dolby B. A Dolby C encoded tape can be played back with a Dolby B decoder without sounding too bad.

The Dolby C system is basically made up of two Dolby B type circuits. It covers frequencies down to about 1000 Hz and improves the S/N ratio by up to 20 dB. Dolby holds that the two-stage compander system used in Dolby C is more accurate than a single compander circuit could be. At no point is the program signal manipulated more than 10 dB by a single circuit.

Other manufacturers have also developed their own NR systems. The *High-Com II* system, developed by Telefunken and Nakamichi, uses a 2:1 compression ratio. (At low input levels, the compression ratio is restored to 1:1.) The S/N ratio is improved by about 20 dB. High-Com II breaks the input signal into two independent bands, each treated separately, to reduce noise "pumping" or "breathing" (audible changes in the noise level). More than two bands would offer an even better improvement in this area, but it would render the system to expensive for most consumer use.

Another popular NR system is *dbx*. This system enjoys quite a bit of popularity because it gives about the most noise reduction now available—about 35 dB.

For a recording studio, an external NR unit is desirable. It offers greater versatility in hooking up the system.

SPEED EFFECTS

A sound can be altered considerably by varying the speed during recording. Played back at the regular speed, the sound takes on a very different character.

The simplest type of speed effect is achieved simply by recording at one speed and playing the tape back at a different speed. For example, if you record at 15 ips and play the tape back at 7 1/2 ips, the tempo is halved, and all notes drop one octave. The tonal characteristics might also be altered because of the change in some of the harmonic relationships. The amplitude envelope also slows down, changing the nature of the sound. The same sort of effects occur if you record at a slow speed and play the tape back at a higher speed.

Most standard tape recorders only offer two or three fixed speeds that are multiples of two. The standard tape speeds increase by doublings:

- 1 7/8 ips
- 3 3/4 ips
- 7 1/2 ips
- 15 ips
- 30 ips

Some machines, however, also have a continuous speed control, allowing the user to fine-tune the running speed. Some very interesting effects can be achieved by changing the speed just slightly—for instance, recording at 7 3/8 ips and playing back at the standard 7 1/2 ips. Even more fascinating effects can be achieved by dynamically varying the speed while recording and playing the tape back at a constant speed.

Dynamic tape speed is generally achieved by applying a variable voltage to the drive motor. If a tape deck has a voltage-controlled type motor, a qualified technician could adapt the circuit to add such a speed control. This is not possible with all types of motors.

REVERSING TAPE DIRECTION

If you're interested in inexpensive, easy-to-produce special effects, try flipping the tape after it's been recorded to play the signal backwards. This reverses the amplitude envelope, as illustrated in Fig. 10-17. Most natural sounds have a longer decay time than attack time. Reversing the tape direction gives a longer attack time than decay time. This makes for some very unusual sounds.

In some cases, one sound source will sound somewhat like a different recognizable sound source. For example, a backwards guitar sounds vaguely like a cross between a violin and organ.

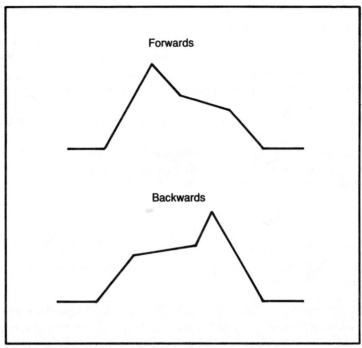

Fig. 10-17. Playing a tape backwards reverses the amplitude envelope, creating novel effects.

ECHO AND REVERBERATION

In some ways, the ideal recording environment might be a room that is completely dead acoustically. An acoustically dead room has no reflected sound waves bouncing off walls or other surfaces that reach the microphone at various times and from various angles. The absence of such echoes makes microphone placement a fairly simple problem. Just point the microphone at the desired sound source and away from any undesired sound source.

The real world isn't quite that cooperative. A completely dead studio is not feasible and, possibly, not even desirable. Such an echoless room is called "dead," because that's the way it sounds. A recording made in this type of environment would sound lifeless and flat. The end result would not be very realistic sounding. Our ears are used to hearing multiple sound reflections bouncing back from various surfaces (Fig. 10-18.) Usually these reflections are not perceived as individual echoes. Instead, they lend a richness and body to the sound. The effect is called *reverberation*.

Too little reverberation sounds dull and unnatural. Too much

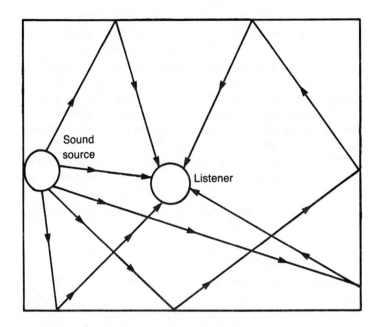

Fig. 10-18. Our ears are accustomed to hearing multiple sound reflections bouncing back from various surfaces in the environment.

reverberation makes the sound muddy and indistinct. In between these extremes are many degrees of reverberation that are useful for different types of recordings. If only for the sake of variety in your recordings, you will probably want to alter the amount of reverberation from time to time.

How can you do this? One approach would be to physically alter the acoustic characteristics of the room. Add absorbing material, such as drapes, carpeting, and overstuffed furniture, to deaden the acoustic environment by reducing sound wave reflections. Similarly, removing these materials increases the liveness or the amount of sound wave reflections. Gobos and sound absorbers may be positioned as needed (see Chapter 6). Heavy drapes over a large window or bare wall can be opened or closed to change the amount of sound wave reflection. Physically rearranging the room and its furnishings also works to alter the acoustic environment and amount of reverberation, but such methods are rather awkward and inconvenient at best. It is hard to be precise with such techniques. You may never be able to recreate a given acoustic environment exactly.

In this section you will learn a few methods for adding artificial reverberation and echo. Before going any further, however, let's define some terms: An echo is a reflected sound wave. Reverberation is many reflected sound waves. These terms are very closely related, but there are important differences. Various technicians distinguish between these terms in different ways, but basically, the reflected sound wave is considered an echo if it is a distinct repetition of the original sound wave. (That is, you hear the sound, then you hear the echo). Reverberation consists of multiple, closely spaced echoes that blend together into a mass of sound. The individual echoes are not discernable.

There is a fairly large gray area between echo and reverberation. Sometimes the reflected sound waves are almost, but not quite distinct. Is this echo or reverberation? It's hard to say. Fortunately, in such cases the terminology is not important, as long as you can get the effect you want.

Artificial Reverberation

There are several ways to add artificial reverberation to a recorded signal. All can sound artificial to a careful listener, but the results can be accurate enough for most purposes. Most artificial reverberators work by repeatedly delaying part of the signal for some brief period to simulate echoes returning from various distances.

One way of delaying the signal is to pass it through a *spring,* as illustrated in Fig. 10-19. The signal passing through the coils of the spring has to travel further than the direct signal. The further the signal has to travel, the longer it will take to reach its destination. The delayed signal is tapped off along various points along the spring, to give multiple delays.

Another approach is the *bucket brigade.* Conceptually this is similar to the old-fashioned fire fighting bucket brigade. A line of

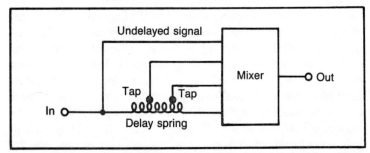

Fig. 10-19. One method of delaying part of a signal for reverberation effects is to pass it through a spring.

people passed buckets of water from the water source to the fire. Each bucket moved from hand to hand, as shown in Fig. 10-20.

The electronic bucket brigade is a series of storage devices (typically capacitances). The signal's instantaneous amplitude is periodically sampled. The first sample is stored in capacitor #1. When the second sample is received, the first sample is passed to capacitor #2, and the second sample is now stored in capacitor #1. This process continues through many stages, each holding the signal for just a few milliseconds. The signal is delayed as long as desired in this manner.

Most artificial reverberation units have at least two controls. *Reverberation rate* delays the time before each echo. *Reverberation depth* is how strong the delayed signals are compared to the original direct signal.

The longer the echo delay, or the slower the reverberation rate, the larger the apparent acoustic environment will be. Larger rooms have longer reverberation times because each echo has to travel further to reach the ear. If the reverberation rate is long enough, it becomes possible to distinguish between the individual echoes.

A common mistake among amateur recordists is to use too much reverberation depth. Every instrument sounds like it's being played at the bottom of a deep well. Such drastic effects can very quickly become quite tiresome to listen to.

In general, artificial reverberation should be used sparingly, especially on low-frequency signals. Bass frequencies tend to become muddied by the multiple echoes. Reverberation should be employed to add life to a sound, not to subtract clarity from it.

Echo Effects

Echo is similar to reverberation except the repetition rate is quite slow, so each repetition is separate and distinct. There may be several repetitions, or there may be just one.

Commercial echo effect devices are available. They function by recording the signal, then playing it back after a brief delay.

It isn't all difficult to create home-brew echo effects. The approach illustrated in Fig. 10-21 works with most three-handed tape decks. Remember that due to the placement of the heads, the signal can be monitored from the playback head a fraction of a second after it is recorded. If you feed this monitor signal back to the record head, it will be rerecorded, creating an echo.

The amount of resistance in the feedback line determines how

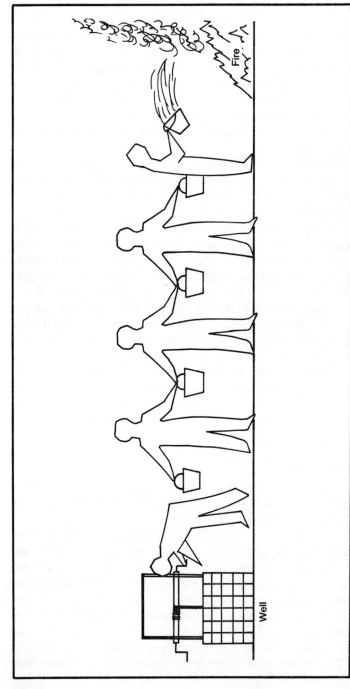

Fig. 10-20. Reverberation effects can also be created by electronically simulating an old-fashioned bucket brigade team.

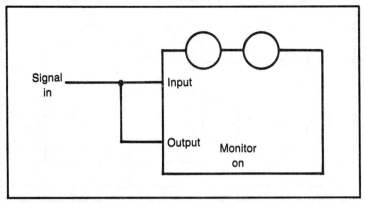

Fig. 10-21. This simple hookup can be used to create basic echo effects on most 3-headed decks.

strong each echo will be and how quickly it dies out. In some cases, with no resistance in the line, you may get feedback oscillations.

An intriguing variation on this approach is shown in Fig. 10-22. This hookup gives you cross-channel echoes. To describe how this effect works, let's keep everything as simple as possible. The original signal is a brief burst of sound recorded on the right channel only. The first echo appears in the left channel. The next echo is in the right channel. The left channel takes the following echo, and so on.

The effect can be quite complex when continuing signals are initially recorded in both channels. This effect works best with fairly simple signals. If you try it on complex, multi-instrument recordings, you may get a muddy blur of indistinct sound. This effect also

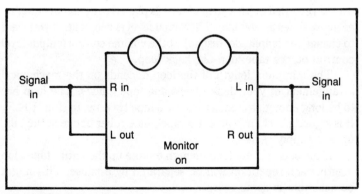

Fig. 10-22. This variation of the hookup shown in Fig. 10-21 produces cross-channel echoes.

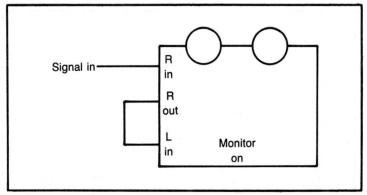

Fig. 10-23. A single cross-channel echo makes a monophonic recording appear to continuously swoop across the stereo picture.

tends to deteriorate the stereo picture, especially with relatively short echo delay times.

For a monophonic recording, you could try a single cross-channel echo (Fig. 10-23). The resulting sound appears to be continuously moving across the stereo picture. While this effect could get monotonous if overused, it can add some interesting "spice" to your recordings now and then.

Tape Loops

Another type of echo-like effect can be achieved with a tape loop. This method is used in mixing. The prepared tape loop is played on one recorder and rerecorded by a second machine.

The tape loop is a relatively short length of tape, with the end spliced to the beginning, making a complete loop (Fig. 10-24). Whatever is recorded on this strip of tape is played over and over again, as many times as you like. The signal level is constant. If you want to change the amplitude, just adjust the volume control (output level control) on the tape loop playback loop.

The minimum length of the loop depends on the mechanical arrangement of your deck's reels and tape guides. The loop has to be long enough to reach the entire tape travel path (Fig. 10-25). It is a good idea to add an extra tape guide roller between the take up and supply reels.

If the sound is not long enough to take up the entire tape loop length, each repetition will be separated by a pause. This might be desirable in some cases.

Longer tape loops are also possible. Simply add more tape

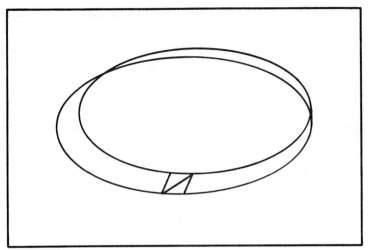

Fig. 10-24. Many interesting special effects can be created with a tape loop.

guides to extend the tape travel path. This is illustrated in Fig. 10-26. All tape guides must be securely mounted and should let the tape travel over them smoothly with as little friction as possible. With a little bit of mechanical ingenuity, you can build a variable length tape loop machine with movable tape guides to alter the size of the loop.

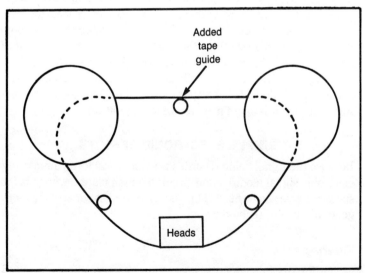

Fig. 10-25. The tape loop has to be long enough to reach the entire tape travel path.

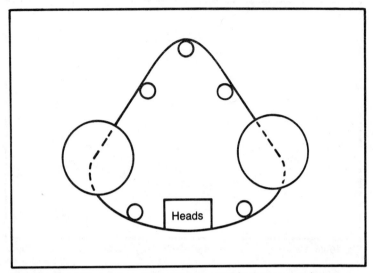

Fig. 10-26. Additional tape guides can be used for longer tape loops.

Changing the speed as the loop is being played can produce some very interesting effects. You could also try running the loop backwards.

By manipulating the playback deck's volume control as the loop is being played, you can get odd effects. For example, one technique is the reverse echo. Ordinarily with repeated echoes, the level starts high and gradually decays into silence. With the tape loop, you can easily reverse this—the echoes gradually build up out of silence to a maximum level. An occasional reverse echo can be a startling effective touch. As with all special effects though, be careful not to overdo it. It is comparable to adding spices while cooking. Too much is far, far worse than none at all.

SIMPLE ELECTRONIC EFFECTS

Because the signal being recorded is in electrical form, almost any electronic signal modification circuit to manipulate the sounds for special effects can be used. This section describes some of the more generally useful techniques.

Filtering

Ordinarily in a sound system, filters and equalizers are used to ensure flat frequency response. For special effects, however, you might occasionally want to deliberately make the frequency response deviate from flat. Filtering can alter the harmonic content of the sound.

Let's consider a tone with the following harmonic components:

140 Hz	fundamental
280 Hz	second harmonic
420 Hz	third harmonic
500 Hz	enharmonic
700 Hz	fifth harmonic
980 Hz	seventh harmonic
1000 Hz	enharmonic
1260 Hz	ninth harmonic

If you pass this signal through a low-pass filter with a cutoff frequency of 650 Hz (for simplicity, assume a perfect filter with an infinitely steep slope), you will get a tone made up of just these frequency components:

140 Hz	fundamental
280 Hz	second harmonic
420 Hz	third harmonic
500 Hz	enharmonic

This will definitely have a different tonal quality than the original sound.

When using fixed filters like this, the fundamental signal shouldn't vary too much because the harmonic content changes with the pitch of the tone and the filter's cutoff frequency remains constant. For instance, if you raise the fundamental frequency to 250 Hz in the example (less than one octave), the harmonic content of the final signal will be stripped down to:

250 Hz	fundamental
500 Hz	second harmonic

Even more drastic effects can be achieved with a band-pass filter. You can even delete the original fundamental, creating a new (probably enharmonic) waveshape.

As an example, use the same original signal from the preceding examples. This time feed the signal through a band-pass filter with a cutoff frequency of 700 Hz and a bandwidth of 300 Hz. In other words, only those frequency components between 400 Hz and

1000 Hz will be passed. In the example, you will be left with:

    420 Hz new apparent fundamental
    500 Hz enharmonic
    700 Hz enharmonic
    980 Hz enharmonic
   1000 Hz enharmonic

Distortion

Usually you want to avoid distortion as much as possible. Sometimes, however, you might want to add some controlled distortion to alter the sound character. The key word here is controlled. You only want certain types of distortion and only in a specific amount.

You could add some distortion effects by altering the amount of bias to the record head, but this is not advisable. It is best to set the record bias level for the best S/N ratio—minimum distortion compromise for the tape you are using and then leave it alone. Use outboard devices to add distortion.

A fuzzbox for an electric guitar can be used. Its purpose is to add distortion to the electric guitar's tone. It can perform the same function on any electrically comparable signal. For best results, the original signal should be kept within the midrange region—that is, tones within the range of a guitar. These are the frequencies the device was designed for. You will have considerably less control over the distortion for very low or very high frequencies.

A super-simple clipping circuit is shown in Fig. 10-27. This device is made up of just a resistor and a zener diode. When the input signal exceeds the zener's breakdown voltage, the excess

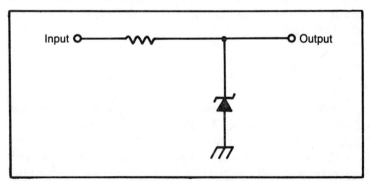

Fig. 10-27. This super-simple clipping circuit can be used to create additional harmonics and a distortion effect.

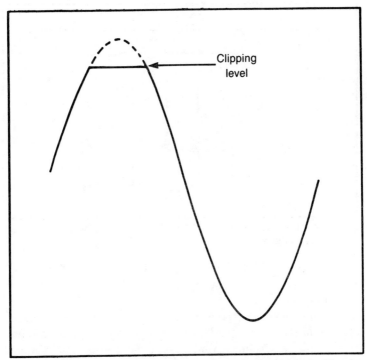

Fig. 10-28. The circuit shown in Fig. 10-27 clips or flattens the tops of the signal's waveforms.

voltage is shunted to ground through the diode. This clips (flattens the top of) the waveform (Fig. 10-28). Clipping will greatly increase the harmonic content of the signal, creating a raspier tone.

You could set up a clipper box with multiple zener diodes that are switch selectable. Each diode has a different breakdown voltage, allowing you to choose the amount of clipping. This idea is illustrated in Fig. 10-29.

Phase Shifting

Phase shifting is a popular effect for rock groups. Dedicated circuits for this purpose, often called *flangers* or *phlangers*, are available in music stores.

The original phlanging effects were created with tape recorders. Try pushing down on the edge of the supply reel during recording, as illustrated in Fig. 10-30. (Be gentle—you don't want to permanently warp the reel or crease the tape). The idea is to continuously vary the amount of pressure on the edge of the reel.

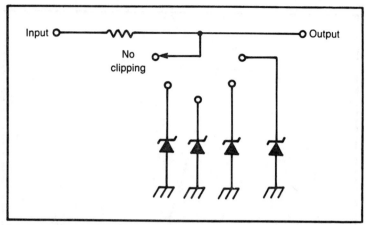

Fig. 10-29. This variation of the circuit shown in Fig. 10-27 allows you to control the clipping level.

This causes the tape to unwind slightly unevenly, resulting in minute variations in the tape speed. The phase of the recorded waveforms will shift about.

Do not slow the tape down so much you can see the speed change during recording. This might damage the recorder's motors and drive system. It would also make the effect too drastic to be pleasing. The speed fluctuations should be very, very slight.

A similar trick is to lightly tap the back of the tape just before it passes the record head (Fig. 10-31). The tapping causes the tape to bounce, changing the distance between the record head and the

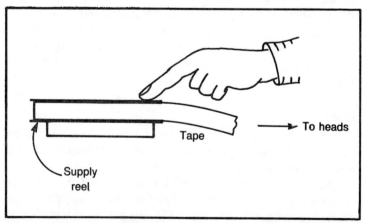

Fig. 10-30. Phlanging effects can be created by applying slight pressure to the rim of the supply reel during recording.

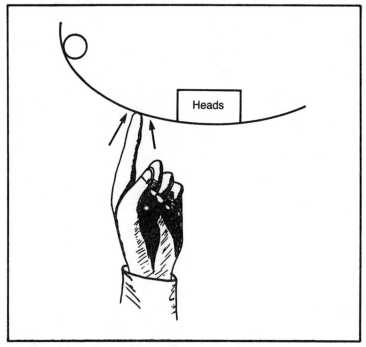

Fig. 10-31. Interesting "warbling" effects can be achieved by lightly tapping the back of the tape just before it passes the record head.

tape. The effect of the magnetic field of the head will be varied by this distance.

Once again, be very gentle. Only a slight amount of tapping motion is required. The effect will be an odd warbling type sound, which could be very interesting in some contexts.

The vital thing to remember when using any special effect is to keep it special. Do not overuse any special effect. What sounds good in moderation can be terribly obnoxious when overdone.

Chapter 11

Tape Types

One of the most important components in your recording system is the magnetic tape itself. All tapes certainly are not created equal. To further complicate matters, the best tape for one system might not be the best choice for a different system.

This chapter examines some of the factors involved in selecting the best tape to use with your recorder. Once you find the best type, stick with it. Using just one brand of tape helps ensure repeatability among your tapes. Other brand tapes might sound somewhat different, especially when special effects are employed.

Take considerable care in selecting the tape you use. Remember, your recording system can only be as good as its weakest component. A poor choice of tape can negate a lot of quality in the equipment's performance.

REFERENCE STANDARDS

Most electrical specifications for recording tape are given in terms of dB. This means some sort of reference tape must be used to define the 0 dB points.

Some manufacturers (unfortunately, not all) use a special, agreed upon reference tape in accordance with DIN (Deutsche Industrie Normen) standards, which were developed in West Germany and are used (not exclusively) throughout the world. The reference tape for this set of standards is the DIN Bezugsband 4.75/3.81 Reference Tape. This is not a particularly exceptional tape, especially in light of recent developments in tape technology, but it is a convenient reference standard for comparisons.

At least six electrical characteristics need to be measured for the unknown tape and compared to the reference tape. These characteristics are:

- ☐ Low-frequency sensitivity
- ☐ High-frequency sensitivity
- ☐ Biased tape noise level
- ☐ Maximum modulation level
- ☐ Distortion level
- ☐ High-frequency saturation

In addition to these electrical characteristics, a number of physical and magnetic characteristics define the quality of the tape in question. All of the electrical characteristic measurements are dependent on the tape deck as well as the tape itself. This makes specifying the quality of a tape a very tricky business at best.

One of the crucial tape deck related factors is the record bias level. This is discussed in more depth later in this chapter. Now, for comparative purposes, a reference bias level and a reference recording level must be set.

A reference recording level is typically set at 25 millimaxwells per millimeter. (A *millimaxwell* is a measure of magnetism. The exact meaning of this term is not important for purposes here.) A midrange tone, typically about 6000 Hz, is recorded at the reference recording level using the reference tape.

Output level varies with the amplitude of the record bias. The bias level is slowly raised past the point where the output level peaks. The output level now starts to drop off as the bias is increased further. The reference bias is the point at which the output level has dropped off 2.5 dB of its peak value. This is illustrated in the graph in Fig. 11-1.

The bias at this point is the reference bias level for the measurements to be made on the tape under test. This is the nominal 0 dB for all comparisons. It is also called the "A" bias.

The entire process is repeated using the tape to be tested. The same input signal and recording level must be used for valid results. The bias level at which the output has dropped 2.5 dB below its peak amplitude is the "B" bias reference. It will be X dB above or below the A (0 dB) bias. Typical results for such a test are illustrated in Fig. 11-2.

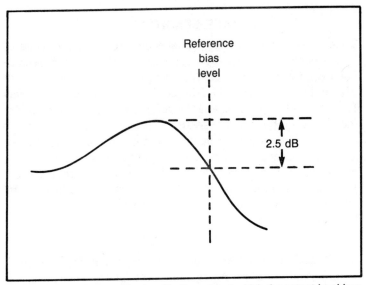

Fig. 11-1. The reference bias level is the point at which the output level has dropped off 2.5 dB below its peak value.

Most electrical measurements to determine the specifications for the tape in question are performed twice—once at reference bias level A and once at reference bias level B.

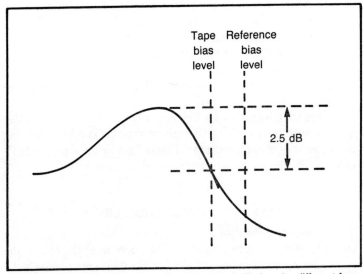

Fig. 11-2. The peak bias level for any given tape is likely to be different from the reference bias level.

TAPE SENSITIVITY

Tape sensitivity is an important specification. It is a measure of how strongly a low-level signal is recorded at the reference recording level. Generally the test recording is made at a level of −20 dB (on the VU meter). The different between the playback levels for the tape being tested and the reference tape (recorded under identical conditions) is the *sensitivity rating* for the tested tape.

Sensitivity ratings are made at both A and B bias and at a low and a high frequency. A tape with a very good sensitivity rating may be labeled "high energy" or "high output." There is no exact level at which these terms may be applied. Don't be influenced by such labels, unless you know the numbers involved.

TAPE NOISE

You'll recall from earlier chapters how important the S/N (signal-to-noise) ratio is for recording equipment. The less noise there is, the better the S/N ratio will be.

S/N ratings are made for the equipment under test. The only noise considered is the noise generated by the equipment itself (usually within the electronic circuitry). Your studio functions as a whole, however, and you are less concerned with the individual ratings for each separate piece of equipment. You want to know how well the system as a whole performs. The important S/N specification for the studio is the combined S/N ratio for all of the equipment in the audio chain—including the tape itself.

Any tape will have some inherent noise level. It is measured with no signal recorded, except for the recording bias. Clearly, the less signal detected upon playback in this test, the better the tape (all other factors being equal).

Since the playback circuitry is involved in the test, the noise is automatically weighted to account for the low-level hearing characteristics of the human ear. This is done by the standard NAB equalization network built into all modern tape decks.

MAXIMUM MODULATION LEVEL

The sensitivity rating for a tape is an indication of the minimum signal level that can be recorded. The opposite of the coin is the maximum modulation level. This specification is not precisely defined. It is generally identified as the maximum recording level that results in no more than 3 percent total third-order harmonic distortion. Spuriously generated harmonics of the recorded signal are highly undesirable, so they should be minimized as much as possible.

Two maximum modulation level measurements are made. Both reference bias levels (A and B) are used for this specification. Maximum modulation level measurements are generally only made with low to midrange test frequencies. The spurious harmonics for a high-frequency test signal would be beyond the audible range, so they wouldn't be of much importance. The reproduction equipment might not be able to reproduce such high frequencies, and even if it could, we couldn't hear them.

Only third-order harmonic distortion (spurious third harmonic generation) is considered in determining the maximum modulation level. Any second-order harmonic distortion (spurious second harmonic generation) that occurs during playback will almost always be due to deck rather than tape problems. A likely culprit is an incorrect waveform for the bias signal. The tape magnetization process will only cause third-order, or perhaps higher-order, harmonic distortion products.

DISTORTION LEVEL

Another distortion-related specification is the distortion level. A medium-low frequency signal (typically about 330 Hz) is recorded, and the level of distortion in the playback signal is measured. The distortion level is a function of bias as well as the tape itself.

Different manufacturers express the distortion level specification in different ways. It may be expressed in percent below the reference recording level or in dB below the reference recording level. Both biases (A and B) are used.

MECHANICAL PROPERTIES

How the tape records the electrical signals is of prime importance. So far, the specifications discussed have dealt with the electrical reproduction, but this is not the whole story. Certain mechanical characteristics of the tape are also important factors in determining the quality of the tape.

How precisely does the width of the tape match its *nominal width* (1/4 inch for reel-to-reel or 1/8 inch for cassette)? The actual tape width might be 0.2425 inch instead of the nominal 0.2500 inch. Large errors in the width affect how evenly the tape passes over the heads.

Closely related to this is the *width tolerance*. Ideally, the tape's width should remain absolutely constant for its entire length. Large variations will cause the tape to jiggle about as it passes by the heads. This can lead to audible problems.

The *yield strength* of the tape is also important. This is usually defined as the amount of force that stretches the tape 5 percent. If the tape is stretched, any signal recorded on it will be speed distorted, some of the magnetic particles might chip off, and the tape is more prone to breakage.

Another mechanical specification for recording tape is the *ultimate breaking strength*—how much force causes the tape to snap. The quality of the *binding material* holding the magnetic particles to the plastic backing is of obvious importance. If the magnetic coating flakes off easily, the tape isn't very good.

Thickness is another important mechanic characteristic. How thick is the plastic backing? How thick is the oxide and how evenly is it applied? And, finally, how thick is the total? Tape thickness decisions are a compromise between economy and tape strength. A thin tape stretches or breaks easier than a thicker tape, but a thicker tape fits onto a reel of a given length, and therefore, the total recording time is lower.

Reel-to-reel tapes are usually supplied in three standard thicknesses:

- ☐ 1.5 mil
- ☐ 1.0 mil
- ☐ 0.5 mil

Compare the recording time for these three tape thicknesses on a 7-inch reel of tape. Assume you are recording in a single direction at a speed of 7 1/2 ips:

1.5 mil	1200 feet	30 minutes
1.0 mil	1800 feet	45 minutes
0.5 mil	2400 feet	60 minutes

For serious recording, you should probably avoid 0.5-mil tape. The possibility of damage to a valuable recording is not worth the tape economy of a too-thin tape—1.0 mil is sufficient for most practical use. For cassette recordings, it is not advisable to use tapes longer than C90. Some experts suggest not using anything longer than C60.

MAGNETIC PROPERTIES

Magnetic characteristics might seem a bit more obscure than the electrical and mechanical properties discussed so far. They are also important, however.

Coercivity, measured in oersteds, is indication of erasure characteristics. This is the field intensity (demagnetizing force) to erase a signal recorded at saturation level, reducing it to zero.

The peak value of 60 Hz AC field applied to tape along its longitudinal direction, which results in at least a 60 dB reduction in a recorded signal, is called the *erasing field*. The required erasing field is also measured in oersteds.

Remanence, stated as X lines per quarter inch, is another aspect of erasure. This specification is a measurement of the lines of flux per unit width of tape when a saturation level recorded signal is reduced to zero (erased). This specification is particularly important. It indicates relative output, distortion, and low-frequency response. In a sense it is a figure of magnetic merit.

Retentivity is quite similar to remanence. This specification, measured in terms of gauss, describes the density or flux per unit cross-sectional area of the tape.

THE RIGHT TAPE FOR YOUR RECORDER

The magnetic tape and the tape recorder form a partnership. Not only do both have to have high quality, they have to be right for each other.

Let's say there are two tapes with comparable specifications. Call them "Acme-45" and "Supersound." Your friend and fellow recordist tells you that Acme-45 is great, but Supersound is lousy. You trust his judgment, but should accept his opinion and not bother trying Supersound? Not unless your recording equipment is identical to his. On your recorder, it is quite possible that Supersound will do a much better job than Acme-45.

Make test recordings on as many different brands of tape as possible. Once you've narrowed the field down to a few good contenders, try purchasing several tapes of each brand from different vendors. Why? So you can get some samples from different batches. Every batch will be slightly different. Some brands have much better consistency from batch to batch than others.

Pick the brand that has the best batch consistency and best results on your recorder. Once you've found the right brand, stick with it. By always using the same tape, you will have better repeatability in your recordings. That gives you better control over the sound—you'll know what to expect.

Frequency Response Tests

If you make your tape comparison tests with recordings of music, try to use the same music for each test recording. Make all of the test recordings under identical conditions. You want to judge the relative sound quality of the tapes, not the music that happens to be recorded on each sample.

You can get better and more reliable comparison results by using certain nonmusical signals in your tests. White noise is a good test signal. You could buy or build a noise generator circuit, but you don't have to. Your stereo system probably already has a noise source.

Tune an FM receiver between stations with the noise muting control off. The hiss you hear is more or less white noise. It's not quite pure white noise. To compensate for transmission losses, certain frequencies are boosted in all modern FM broadcasts. FM receivers have complimentary de-emphasis circuits to compensate for the boost. The de-emphasis circuitry affects the frequency response of the between-station hiss.

Fortunately, you are making comparative, rather than absolute, measurements. Because the same equalization is applied to the signal for all of the tests, it doesn't really matter for your purposes here.

The first step is to hook up the tape deck into the tape monitor circuit. The original or the recorded signal can be selected via the tape monitor switch. The idea is to switch back and forth between the two and listen for any differences.

The ear has a somewhat different frequency response at different volumes, so it is very important that both the original and the recorded signal be at *exactly* the same volume. If the amplitudes of the two signals are even slightly off, you might not get meaningful results from the listening tests.

Most tape decks have an output level control so that the receiver's or amplifier's volume control can be left at a single setting. The tape deck's output level is adjusted to match the original signal.

The best way to match the signal levels is to attach a wideband AC voltmeter to the speaker output terminals of the receiver or amplifier. Adjust the tape deck's output level control so that you get the same meter reading for both positions of the tape monitor switch.

If you have a three-headed deck, set it up to record with the monitor (playback head) on. Alternatively, if you just have a two-headed deck, record several minutes of the between station hiss, rewind the tape, and play it back for the tests.

Listen very carefully while switching back and forth between the original signal and the taped signal. Even small differences in the frequency responses will be audible.

Distortion Tests

The frequency response of your tape is certainly important, but so is distortion. The bias setting is always a compromise between frequency response and distortion. Good frequency response can be achieved with almost any tape if distortion is ignored, but the audio results are likely to be quite unsatisfactory.

To test for distortion, perform a series of tests like those described in the preceding section. Instead of using white noise (FM between station hiss) as your signal source, use an audio oscillator to generate a continuous tone. If you don't have an oscillator circuit available, you can easily build the circuit shown in Fig. 11-3. The output level of the oscillator should be 1 volt.

The frequency of the test tone should be somewhere between 500 and 1000 Hz. Connect an AC voltmeter with a dB scale to the output terminals of the tape deck. If your AC voltmeter does not have a dB calibrated scale, you can use the following chart to convert voltage ratios to dB:

Measured voltage	*+ dB*
1.000 × ref	0
1.122 × ref	1
1.259 × ref	2
1.413 × ref	3
1.585 × ref	4
1.778 × ref	5
1.995 × ref	6
2.239 × ref	7
2.512 × ref	8
2.818 × ref	9
3.162 × ref	10

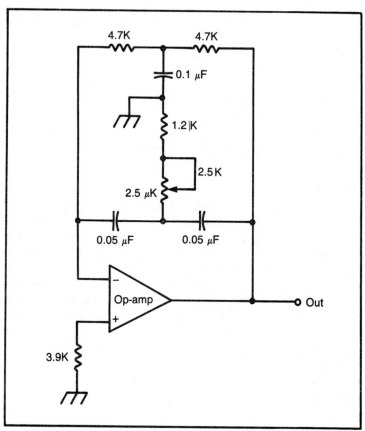

Fig. 11-3. This simple oscillator circuit can be used as a signal source for the distortion tests.

To set up your reference level, set up the tape deck to record at a 0 dB level as indicated on the VU meters. Notice the reading on the AC voltmeter at this setting. If possible, adjust the voltmeter so it is also indicating 0 dB, or some convenient reference point.

The AC voltmeter is now set up to function as an external pseudo-VU meter. You need the external meter because these tests almost definitely go beyond the full scale range of your deck's built-in meters. In this test you will deliberately overrecord the tape. Do not be concerned about pegging the built-in VU meter's pointer. No damage will be done.

With a three-headed tape deck, listen to the monitor (immediate playback after recording) as you gradually increase the level of the signal being recorded in 1 dB increments. Compare the purity of the recorded tone with the original signal. As you increase the level setting, you will eventually hear a clear increase in the amount of distortion. This is the point where the distortion is approximately 3 percent.

The exact amount of distortion is not really the issue here. If you are consistent about what the "unacceptable" level of distortion is for each tape you test, you will be able to make valid comparisons between the various tapes to find the best performance with regard to distortion. The tape (or tapes) that allow the highest level setting before the distortion becomes objectionable is the winner(s) of this set of tests.

If you have a two-headed tape deck, which does not allow immediate monitoring, the procedure is slightly complicated. Record the signal at 0 dB for five seconds, increase the level to +1 dB for five seconds, then to +2 dB, and so forth, up to +10 dB in 1 dB steps of five seconds each.

When you play back the tape, keep track of the timing for each increase in level and listen for the perceptible increase in the distortion. It's a little trickier to determine the dB setting this way, but if you work slowly and keep track of how many step increases have occurred, you shouldn't have any problem.

S/N Ratio Tests

It should be clear by now that the S/N (signal-to-noise) ratio is an important specification in audio work. The S/N ratio defines the dynamic range (extend of usable amplitudes). The minimum usable amplitude is set by the level of the noise floor. Weaker signals can be recorded, but they will be lost beneath the tape hiss. The maximum usable amplitude limit is set by the point at which distortion sets in.

The wider the dynamic range, the better and more realistic recordings you can make. The maximum usable recording level is generally assumed to be the point at which 0 dB is indicated on the VU meter. Assuming that the distortion level is constant, the dynamic range can be increased by lowering the noise level, that is, the better the S/N ratio, the better the dynamic range.

Reputable manufacturers usually specify the S/N ratio of their product(s). You should be able to find the S/N ratio of your tape deck, for example, somewhere in the owner's manual.

In the studio, however, you are more concerned with the S/N ratio of the system as a whole—the tape deck, **the tape** itself, the amplifier, the speakers, the connecting cables, any mixers or special effect circuits in the line—in short, the combined total of everything that carries the signal in any way contributes to some degree to the overall noise level. Unfortunately, you can't just add together the S/N figures for the various separate components. The only way to determine the actual S/N ratio of your complete system is to measure it yourself.

The tape itself probably is the biggest contributer to the overall system noise. Luckily, it is also the component that is easiest to change—just buy a different brand of tape next time. The following test can be used to determine the best tape for your system in terms of the S/N ratio.

To find the noise floor of your recorder and the tape being tested, connect an AC voltmeter to the output terminals of the tape deck. A very sensitive meter must be used here because the noise signal is quite small. Typically the noise level is approximately 1 mV (0.001 volt).

If you intend to use a noise reduction system when you make your actual recordings, perform this test with the noise reduction circuit on. Record a minute or so of tape with all input level controls turned all the way down and with the input terminals of the tape deck shorted out. Rewind the tape and measure the output level during playback. The only signal present on the tape will be the noise floor hiss.

Do not try to make the measurement by monitoring the playback head during recording with a three-headed deck. The recording bias signal will get through to the output and invalidate the voltmeter's reading.

The results can be given in terms of dB with the same 0 dB reference point used in the distortion tests. Some typical voltage to dB conversions are:

Measured voltage	*dB*
0.00100 × ref	−60
0.00126 × ref	−58
0.00158 × ref	−56
0.00200 × ref	−54
0.00251 × ref	−52
0.00316 × ref	−50
0.00398 × ref	−48
0.00501 × ref	−46

0.00631 × ref	−44
0.00798 × ref	−42
0.01000 × ref	−40

This is the minimum usable recording level. It can be combined with the maximum usable recording level found in the distortion tests to find the full dynamic range of the tape in question.

As an example, let's say the 0 dB point on the voltmeter was 2.2 volts. When we tested Supersound tape in the distortion test, noticeable distortion set in when the voltmeter reading was about 4.9 volts, which corresponds to approximately +7 dB. The noise test produced a voltage reading of about 4.4 mV (0.0044 volt), which is about −54 dB. Therefore, the dynamic range of our hypothetical Supersound tape runs from approximately −54 dB to about +7 dB, giving a full range of just over 60 dB. This is a pretty good range.

Making the Selection

Inevitably, one tape has the best frequency response, while a second tape has the best distortion figures, and yet a third tape has the lowest noise floor. Compromise is inescapable. You want to choose the tape with the best average of all three of these specifications. The decision is never easy, but these tests will, at least, help you weed out some of the candidates.

Make some musical recordings and listen with a critical ear. You might find that you like the "sound" of one particular tape. Pick one brand of tape that works well with your system and try to use it exclusively. This is the best way to ensure top performance from your recording system at all times. As you get more and more familiar with the specific characteristics of the tape you use, you can better compensate for any minor defects it maybe exhibit.

One warning however—manufacturers occasionally change the formulation of their tapes slightly from time to time, without any kind of notice. There may be no difference in the packaging at all. For this reason, and because your system components will eventually age and change their operating characteristics slightly, it is probably a good idea to repeat these tests periodically to make sure you're still using the best tape for your system.

Remember, there is no one best tape. Anybody that tries to rank various brands of tape in a generalized fashion is overlooking a lot of important variables. Unless you have the exact same equipment, that was used in the tests and should be "The Ten Best Tapes," is completely worthless and should be ignored. There is no *best* tape. All you can do is find the best tape for your specific system, and you'll probably have to do that yourself.

Some tape deck manufacturers recommend certain brands of tape in their owner's manuals. That gives you a head start, but don't even take their word for it. As stated earlier, tape formulations are sometimes changed. The deck manufacturer might have missed a brand that gives superior performance on that deck. Other components in your system (which the tape deck manufacturer can't account for) and personal taste in where to make the necessary compromises can also be factors in selecting the best tape for the system.

Whether you perform the technical tests described in this chapter, or just rely on extensive listening tests, to get the best performance out of your recording system you must experiment with as many different brands of tape as you can find. The most expensive brand available may not be the best for your individual needs.

WATCH OUT FOR BARGAINS

I strongly advise that you stick with familiar name brand tapes. Especially beware of white box (no brand name) specials. These can be very cheap, and might be a bargain, but are probably a waste of money. Usually no-label tapes are a name manufacturer's rejects or tapes made without adequate quality control. Sometimes they are previously used tapes and might be full of poorly made splices. You get what you pay for, and this kind of bargain is usually no bargain.

"El cheapie" tapes should especially be avoided if you are working with cassettes. The cassette housing must be precisely manufactured to function properly. Those tapes sold in discount stores in bags of three for a dollar are almost guaranteed losers. Decent tape and housings simply can't be manufactured for that price. Quality control is probably nonexistent.

I know of one person who bought four packages of three "bargain" C60 cassettes. Out of the twelve tapes, two jammed when he first tried to use them. Four ran when he recorded them, but jammed the first time he tried to play them back. Three jammed within four or five playings. One ran, but wobbled so much that anything recorded on it was completely unlistenable. Even the two tapes that worked mechanically had extremely poor sound quality. In short, my friend threw away money.

Sometimes these bargain tapes are excessively abrasive and can quicken head wear. Jammed cassettes can, in some cases, actually damage the drive mechanism of the tape deck itself. It doesn't make much sense to risk a $50 to $100 repair bill to save a dollar or two on tape. Especially since the odds are very good the cheap tape won't give decent performance even if it does work.

If a bargain price looks too good to be true, it probably is. Do yourself a big favor and pass up such bargains without a moment's hesitation.

Name brand tape manufacturers know they rely on repeat customers to stay in business. Consequently, they do everything they can to ensure that their products give good performance. Most manufacturers and audio stores are quite reasonable about replacing defective tapes.

Bear in mind though, if you had a valuable, one-of-a-kind recording on the tape before it jammed, or whatever, the manufacturer can only replace the blank tape. The lost recording is your problem. This is why it is so important to use quality tape that you can trust with reasonable assurance. A jammed or broken tape, or one with an unacceptable number of dropouts, can be a disaster in a live recording situation.

TAPE FORMULATIONS

The primary factor in creating differences between various tapes is the formulation of the magnetic material that is applied to the plastic base. The magnetic layer of a tape consists of some type of magnetically-sensitive metallic oxide particles. These particles are microscopic and are dispersed in a binder substance that holds them to the tape. This is the material that does the actual recording, so it's importance is obvious.

Because of the importance of the tape formulation, most manufacturers are careful to keep the exact formulas they use secret. The exact formulation used is responsible for the characteristic sound of each brand of tape. Many variations in the formulation may cause differences between different batches of the same brand tape. The better manufacturers do their best to ensure batch-to-batch consistency. This is another good reason to avoid bargain brands. The recording characteristics might vary considerably from batch to batch.

Several different types of materials have been used in tape formulations. This is especially true of cassettes. A lot of work has been done to find a tape formulation that will compensate for the inherent limitations of this slow speed, narrow track width medium.

The two most common classes of magnetic materials used are ferric oxides and chromium dioxide.

Ferric Oxide Tapes

Until a few years ago, all tapes used a ferric-oxide-based formulation. Ferric oxide (Fe_2O_3) is similar to iron rust.

Ferric oxide tapes tend to have a well-balanced frequency response. The reproduced sound is rather smooth, which is pleasant to listen to.

A standard equalization curve is used for most ferric oxide tapes. The time constant of the RC network is 120 uS (microseconds). When this was the only tape formulation available, all tape decks could have a standardized equalization network, and the user wouldn't have to worry about it. Some newer tape formulations require different equalization curves, so switch-selectable equalization is becoming more and more common on tape decks, especially cassette decks for which most nonferric tape formulations were devised.

Chromium Dioxide Tapes

One of the main limitations of recording, especially for the cassette, is the high-frequency response. High frequencies tend to be the most blocked by hiss (tape noise) and are the weakest recorded signals—which is true for all tape formats. Cassettes have additional limitations because of their low speed (1 7/8 ips) and the narrow track width. Both of these factors also limit high-frequency response.

Chromium dioxide (CrO_2) tape formulations were developed to help combat these problems. While some reel-to-reel chromium dioxide tapes have been marketed, this formulation, along with most other nonferric oxide formulations, is used primarily for cassette tapes.

Chromium dioxide tapes record strong high-frequency signals, although they are somewhat weaker than ferric oxide tapes at the lower end of the audio spectrum. The treble region is about 4.5 dB stronger than for a ferric oxide tape. The equalization network needs to apply less boost to the high end of the playback signal. Tape noise and hiss are equalized right along with the recorded signal, so the result is less high-frequency hiss in playback. Since tape noise is most objectionable at high frequencies, chromium dioxide can be set to have approximately a 4.5 dB reduction in noise, compared to ferric oxide tapes.

Overall, the sound of chromium dioxide recordings is rather bright with a lot of "punch." This tape formulation is well suited for many rock and jazz recordings, but the uneven frequency response may be undesirable for classic recordings. The equalization network for chromium dioxide tapes has a time constant of 70 uS (microseconds).

Other Tape Formulations

Ferrichrome tapes were developed in an attempt to get the best of both worlds. A thin chromium dioxide layer is placed atop a slightly thicker ferric oxide layer. The surface chromium layer handles the high frequencies, while the lower ferric layer takes care of the low frequencies. This combines the advantages of both formulations. The major disadvantage is cost. Ferrichrome tapes tend to be quite expensive because of the required added steps in manufacturing. The equalization curve for ferrichrome tapes has not been standardized. Some use the standard ferric setting, and others work better with the chrome setting.

Another compromise formulation is called *ferricobalt*. Only a single layer of recording material is applied to the plastic backing. Each microscopic ferric molecule is bonded to a cobalt molecule. High-frequency sensitivity is improved by this process. The S/N ratio is also better with ferricobalt tapes, widening the dynamic range. Once again, ferricobalt tapes tend to be quite expensive. The chemical processes involved in bonding the cobalt molecules to the ferric particles are complex. Ferricobalt tapes are sometimes called *CrO_2 equivalent ferrics* because they use the same equalization curve as chromium dioxide tapes.

Another fairly new type of tape formulation are the so-called "metal" tapes. Most tape formulations employ microscopic particles of metallic oxide to record the signal. The new type of tape uses fine metal (not metal oxide) particles. Both low-frequency and high-frequency signals are increased for metal tapes. The maximum output from a metal tape can be 5 to 10 dB greater than for a chromium dioxide tape, or 3 to 7 dB greater than for a high-quality ferric oxide tape. This higher output translates to a better S/N ratio and improved dynamic range. There is also less distortion because the tape is less likely to be overloaded.

Metal tapes can be played back with chromium dioxide (70 uS) equalization. Any deck that can handle chromium dioxide tapes can play back metal tapes. Recording is another story. The record bias current must be considerably greater to record a metal tape than any other tape formulation. To erase a metal tape, a much higher current must be applied to the erase head than for other types of tape. Many current record and erase heads can't handle the high currents required for recording metal tapes. A deck must be designed with metal capability in mind to record this new formulation.

SUMMARY

You can spend a small fortune on your recording equipment, and still not get high-quality recording—unless you are careful in selecting the tape you use. In many cases, you will get better results with high-quality tape on moderately priced equipment than you will with mediocre tape on the most expensive recording equipment you can find. The tape itself is an extremely important part of the recording chain.

Index

Index

A
accessories, 153
acoustics, 88
  introduction to, 1
amplifiers, 75
amplitude, 6
attack time, 129
average volume, 6
axes, 8

B
band reject filter, 157
band-pass filter, 157
basstraps, 95
bel (B), 10
bias, 39, 188
  peak, 189
  reference, 189
bidirectional microphone, 63
bidirectional monaural recording, 29
bidirectional recorders, 41
bidirectional stereophonic recording, 29
boosting, 38

C
capacitor, 60
carbon microphone, 55, 56
cardiod microphones, 63
cassette tape
  format of, 46
cassette tape recorders, 42
center frequency, 157
ceramic microphone, 58
chromium dioxide tapes, 202
clipping circuit, 182
clipping distortion, 74
coercivity, 193
coloration, 95
compander noise reduction system, 166
complimentary expander, 168
compression, 167
condenser microphone, 60
console equalizer, 160-162, 164
crystal microphone, 57
cutoff frequency, 154
cycles per second (cps), 11

D
dBm scale, 52
dead environment, 101
"dead recording", 94
decay time, 129, 130
decibels (dB), 10, 51
  chart of sound, 52
demagnetization, 114
Deutsche Industrie Normen (DIN) tape standard, 187
dielectric, 60
distortion, 54, 182, 191
Dolby noise reduction systems, 69, 169
dynamic microphone, 58

E

echo, 172
  cross-channel, 177
  effects of, 175
  tape loop, 178
eight-tracks, 47
  divisions of, 48
Elcassette, 48
electret microphones, 60
enharmonic, 14, 18
equalization, 35, 37, 153
equalization boosting, 38
erase head, 27

F

ferric oxide tapes, 202
ferrichrome tapes, 203
ferricobalt tapes, 203
filtering, 153, 180
filters, 153
  band reject, 157, 158
  band-pass, 157, 158
  high-pass, 54, 156
  low-pass, 20, 154
  notch, 157
  op amp, 156
  Q in, 157
  RC (resistor-capacitor) network, 155
flangers, 183
four-channel recorder, 70
frequencies, 11
  standard western music scale, 12
frequency response, 78, 194
  factors determining, 31
fundamental frequency, 14, 16
fuzzbox, 182

G

gobo isolator panel, 101
graphic equalizers, 159, 161

H

harmonics, 13, 14
head cleaning, 111
head cleaning cassette, 113
headphones, 84
hearing
  nature of, 1
  threshold of human, 51
hertz (Hz), 11
Hi-Com II noise reduction system, 170
high-amplitude signal, 6
high-frequency roll-off, 33
high-pass filter, 54, 156

I

impedance, 76
  calculations for, 87
  matching, 84
  ratings, 76
instantaneous amplitude, 6, 8, 26
insulator, 60
internal oscillator, 40
isolation systems, 98
isolator panel (gobo), 101

K

kilohertz (kHz), 11

L

live environment, 101
live recording, 94
  equipment for, 103
  setting up for, 105
  test recordings for, 106
logic control, 72
low-amplitude signal, 6
low-pass filter, 20, 154
lubrication, 116

M

magnetic flux, 35
maximum load impedance, 86
megahertz (MHz), 11
metal tapes, 203
microphones
  bidirectional, 63
  carbon, 55, 56
  cardiod, 63
  ceramic, 58
  condenser, 60
  crystal, 57
  dynamic, 58, 59
  electret, 60
  omnidirectional, 61
  pickup patterns of, 61
  ribbon, 59
  schematic symbol for, 56
  unidirectional, 61
millimaxwell, 188
minimum load impedance, 85
mixers, 148
  pots in, 150
  simple passive, 149
mixing
  editing of, 152
  equipment used in, 148
  four-channel decks and, 143
  multideck approach to, 145
  multideck synchronization, 146
  overdubbing in, 150

sound-on-sound, 139
sound-with-sound, 140
track bouncing and, 144
modulation levels, 190
monitor system for recording studios, 72

N

negative maximum level, 8
noise
 generator of, 21
 pink, 20
 white, 20
noise reduction systems, 69, 165
 compander, 166
 Dolby, 169
 dual-ended, 166
 single-ended, 166
nominal width, 191
notch filter, 157
null position, 2

O

octaves, 12
omnidirectional microphone, 61
op amp, 156
Oriental scale, 13
overdubbing, 150
overtones, 14, 18

P

parametric equalizer, 163
peak bias, 189
peak filtering, 162
pentatonic scale, 13
phase shifting, 183
piezoelectric effect, 57
pink noise, 20
pitches, 13
playback head, 27
 typical voltage vs. frequency graph for, 36
potentiometers, 159, 163
pots, 150
power handling, 77
power levels, calculation of, 52
presence control, 157
print-through problems, 110

Q

quadrophonic (four-channel) recording, 29
quarter-track stereophonic recording, 29

R

RC network, 155
record bias, 39
 amplitude of, 40
 frequency of, 40
record head, 23, 26, 27
recording studio
 acoustics in, 88
 headphones for, 84
 impedance matching for, 84
 monitor system for, 72
 power requirements of, 73
 room acoustics for, 82, 94
 selecting tape machine for, 67-72
 selection of amplifier for, 75
 selection of speakers for, 76
 soundproofing in, 88-94
recording tape
 alignment of particles on, 26
 binding material of, 192
 care of, 109
 distortion level of, 191
 distortion test for, 195
 formulations of, 23, 187, 201
 frequency response test for, 194
 handling of, 110
 magnetic field effect on, 25
 magnetic properties of, 193
 makeup of, 24
 maximum modulation level of, 190
 mechanical properties of, 191
 noise rating of, 190
 nominal width, 191
 reference standards for, 187
 S/N ratio test for, 197
 selection of, 193-201
 sensitivity rating of, 190
 storage of, 109
 thickness of, 192
 ultimate breaking strength of, 192
 width tolerance of, 192
 yield strength of, 192
reel-to-reel recorders, 41
reference bias, 189
release time, 129
remanence, 193
resonators, 95, 97
rest or null position, 2
retentivity, 193
reverberation, 101, 172
 artificial, 174
 depth of, 175
 rate of, 175
reversing tape direction, 171
ribbon microphone, 59
roll-off, high frequency, 33

room acoustics, 94
  basstraps in, 95
  breaking up standing waves in, 95
  coloration in, 95
  dynamic control in, 101
  floating platforms in, 97
  floors in, 96
  gobo isolator panel in, 101
  isolation systems for, 98
  microphone placement in, 97
  reverberation in, 101
  tent in, 98
root mean square (RMS) average, 9

S

S/N specifications, 54
second harmonic, 15
shelf filtering, 163
signal voltage, 27
signal-to-noise ratio (S/N), 54
sine wave, 8, 9
single-track monaural recording, 29
sound
  air pressure and, 6
  compression of air to produce, 3
  nature of, 1
  vibration to produce, 2
  waveform axes of, 8
  waveforms of, 5
sound envelopes, 131
sound trap, 89
sound waves, 5
sound-on-sound mixing, 139
sound-with-sound mixing, 140
soundproofing, 88-94
speakers, 76
  frequency response of, 78-83
  impedance ratings of, 76
  power handling of, 77
  woofer in, 83
special effects, 153
spectrum analyzer, 165
speed effects, 170
splicing
  backward sounds, 134
  basic method for, 121
  blocks for, 126
  broken tape, 122
  changing sound envelopes by, 132
  crossover, 136
  incomplete sounds, 128
  jigsaw, 136
  reasons for, 119
  two different sounds, 135
square wave, 17

stereo recording, 31
subsonic frequencies, 12
sustain time, 129, 131
sync switch, 143

T

tape loops, 178
tape recorders
  care of, 109
  cassette tape format, 32, 42
  choosing correct format for, 48
  cleaning equipment and supplies for, 111
  demagnetization of, 114
  8-track cartridge format, 47
  equalization in, 35
  erase head of, 27
  factors determining frequency response in, 31
  head arrangement of, 28
  how signals are recorded by, 23
  lubrication of, 116
  maintenance and repairs of, 116
  playback head of, 27
  record bias in, 39
  reel-to-reel tape format, 31, 41
  tape formats for, 41
  tape speed in, 33
  tape tracks for, 29
  voltage signal in, 34
tape speed, 33
tape tracks, 29
tent, 98
threshold of hearing, 51
tone controls, 157
tones, 13
track bouncing, 144
triangle wave, 18
two-track monaural recording, 29
two-track stereophonic recording, 29

U

ultrasonic frequencies, 12
undertones, 14, 18
unidirectional microphone, 61

V

vibration cycle, 6
voltage signal, 34
volume units (VU), 53
VU meters, 53

W

warbling effects, 185
weighting curve, 54
white noise, 20

Other Bestsellers From TAB

☐ **DEMYSTIFYING COMPACT DISCS: A GUIDE TO DIGITAL AUDIO—Sweeney**

This time- and money-saving sourcebook will give you the background knowledge you need to find the CD player and accessories that meet your needs for the best possible price. The strengths and weaknesses of this new medium are carefully examined along with tips on how to get the best performance, available accessories and even a look at the future of digital audio and optical discs for computers. 176 pp., 55 illus.
**Paper $9.95   Hard $15.95
Book No. 2728**

☐ **SUCCESSFUL SOUND SYSTEM OPERATION—Everest**

The key to successful audio sound is not the microphone, the amplifiers, the loudspeaker, or even the acoustics . . . it's you the sound system operator. Here's an invaluable guide, written by a professional acoustics consultant and audio engineer, that focuses on how you can effectively coordinate every component in your sound system for maximum performance. 336 pp., 321 illus. 7″ × 10″.
Paper $17.95   Book No. 2606

☐ **DESIGNING, BUILDING AND TESTING YOUR OWN SPEAKER SYSTEM . . . WITH PROJECTS —2nd Edition—Weems**

You can have a stereo or hi-fi speaker system that rivals the most expensive units on today's market . . . *at a fraction of the ready-made cost!* *Everything* you need to get started is right here in this completely revised sourcebook that includes everything you need to know about designing, building, and testing every aspect of a first-class speaker system. 192 pp., 152 illus.
Paper $10.95   Book No. 1964

☐ **MICROPHONES—3rd Edition—Clifford**

Here's all the information you need to choose the proper microphone and operate it effectively for truly professional quality sound, no matter what your recording job . . . indoors or out! Get better stereo, achieve special effects, make a speaker clearly intelligible to an audience . . . record almost anything from a symphony to a child's voice, from a rock group to a soprano soloist, from a business seminar to a political rally! 352 pp., 235 illus.
**Paper $16.95   Hard $22.95
Book No. 2675**

☐ **AUDIO SWEETENING FOR FILM AND TV—Hubatka, Hull, and Sanders**

Contains the information you need to create high quality audio for film and video productions. It provides complete instructions for producers who want to improve video tracks, and experienced audio engineers who want to learn video editing techniques! Shows how to create sound tracks using high quality recorders, monitors, amplifiers, mixing consoles, and synchronizers. 240 pp., 99 illus. 6″ × 9″.
Hard $30.00   Book No. 1994

☐ **AM STEREO AND TV STEREO—NEW SOUND DIMENSIONS**

Here's a look at the new sound in AM radio and TV broadcasting that gives much needed advice on equipment availability and operation and provides insight into FCC regulations and equipment specifications. Here's where you will discover the latest information on all types of AM stereo receivers and transmitters. Plus, get a complete introduction to Multi-Channel TV sound! 192 pp., 127 illus. 7″ × 10″.
**Paper $12.95   Hard $17.95
Book No. 1932**

Other Bestsellers From TAB

☐ **VIDEO PRODUCTION THE PROFESSIONAL WAY—CAIATI**

Now, right in step with the latest techniques and equipment available to the video recording world, this hands-on sourcebook provides an in-depth introduction to professional-quality video tape production that no video enthusiast can afford to miss! Caiati provides guidance in all the basic taping and editing procedures, plus tips on equipment maintenance and troubleshooting. 256 pp., 306 illus. 7" × 10".
**Paper $16.95   Hard $24.95
Book No. 1915**

☐ **ACOUSTIC TECHNIQUES FOR HOME AND STUDIO—2nd Edition—Everest**

This is an essential sourcebook for every dedicated audio enthusiast! The completely updated edition of a classic reference, it provides pro guidance on creating the best possible sound recording and listening environments. Covering everything from how sound is transmitted to specific design in applications and actual photos of working hobby and professional studios from around the world. 352 pp., 271 illus.
Paper $15.95   Book No. 1696

☐ **BUYER'S GUIDE TO COMPONENT TV**

An essential sourcebook for anyone contemplating the purchase of new TV or home video equipment! Fully covers monitors and projection screen TVs, stereo TV sound units, VCRs and video cameras, even home computers and satellite TV dishes! Provides realistic advice on how and where to find the best buys on various components, plus valuable tips on how to get more professional results from your video units. 224 pp., 177 illus. 7" × 10".
**Paper $12.95   Hard $19.95
Book No. 1881**

☐ **DIGITAL AUDIO TECHNOLOGY**

Here is a firsthand report on the latest developments in the science of audio recording and playback—the combination of two modern technologies: digital electronics—as used in computers and satellite communications, *and* audio. PCM is already being used in putting together master tapes and its details are revealed in this book—must reading for the audiophile and computer buff alike! 320 pp., 210 illus.
Paper $14.95   Book No. 1451

*Prices subject to change without notice.

Look for these and other TAB books at your local bookstore.

TAB BOOKS Inc.
P.O. Box 40
Blue Ridge Summit, PA 17214

Send for FREE TAB catalog describing over 1200 current titles in print.